Adventures of an Inventor

or

How to Survive and Succeed in Inventing Business

Adventures of an Inventor

or

How to Survive and Succeed
in Inventing Business

by
Jacob Fraden

San Diego, California

Adventures of an Inventor

or

How to Survive and Succeed in Inventing Business

by Jacob Fraden

Published by:

Post Office Box 927412

San Diego, CA 92192-7412 U.S.A.

All rights reserved. No part of this book may be reproduced or transmitted in any form or by any means, electronic or mechanical, including photocopying, recording or by any information storage and retrieval system without written permission from the author, except for the inclusion of brief quotation in a review.

Copyright © 1996 by Jacob Fraden

Library of Congress Catalog Card Number: 95-82166

Fraden, Jacob

Adventures of an inventor or how to survive and succeed in inventing business/ Jacob Fraden

384 p.

ISBN   0-9649927-3-6

10 9 8 7 6 5 4 3 2 1

For the book ordering information turn to the last page

To those few who don't give up

CONTENTS

Preface	**9**
Chapter 1	
Making Movies	**11**
Chapter 2	
The First American	**33**
Chapter 3	
The Mind Readers	**49**
Chapter 4	
Making an Inventor	**71**
Chapter 5	
Escape from the P.O. Box	**95**
Chapter 6	
Spooky Business	**127**
Chapter 7	
Starting Over	**143**
Chapter 8	
Moonlighting	**181**
Chapter 9	
Up the Creek without a Puddle	**187**
Chapter 10	
Fight for Light	**221**
Chapter 11	
High Pressure	**233**
Chapter 12	
Mikie	**259**
Chapter 13	
Instant Thermometer	**285**
Appendix:	
Making Inventions	**343**

Preface

This book is not the autobiography of a movie star, a politician, or any other celebrity. My memoirs will be of no interest to the reader looking for unfamiliar facts in the lives of familiar personalities. Nor is this book a manual on making inventions or running a small business. What, then, is it and why do I think you should read it?

I was lucky to have two lives—one before I came to America from the Soviet Union, and one after. Probably because of the adventurous nature of my character, no matter which side of the Iron Curtain my home was on, I was constantly searching for novelties and adventures. And I was fortunate to have quite a few of these in nearly everything I did, from making movies to inventing high-tech electronic devices. Or perhaps, I just perceived many events in life as adventurous. But does it make any difference?

This book is about a lifelong quest for new ideas and bringing innovative products to reality. A search for newness is a natural and spontaneous process, pleasant and challenging, almost a child play. Alas, turning an idea into a real product is a monumental and often disappointing task. It requires persistence, resilience and most of all—an adventurous character.

A socialistic system made the inventing business nearly impossible and very much frustrating, where one could spend a lifetime beating that system with a very small chance to succeed. Unfortunately, I found that a capitalistic system is not well geared for the innovative process either, yet it still offers some opportunities. So every new idea is a starting point in a quest for a dream.

It's like a game where one needs a great deal of luck. And if you want to win the game (who doesn't?), you better figure out how to beat the system. I have learned something on my road to success and wish to pass on a few bits of experience to others who might use them. Why not avoid pitfalls if you know that someone has already fallen there? Indeed, it's always safer to learn from someone else's experience. Of course, experience won't safeguard anyone from making mistakes, but at least it may help one to recognize them sooner.

Modern-day adventures are not what they used to be. They are of a more civilized nature, so to speak, not like those of Robinson Crusoe or the pirates of the Caribbean. Yet though they may well be more prosaic on the surface, inwardly there is still high risk, tense emotion, and bubbling ardency. As it was said quite correctly, the road to a good invention is paved with hell.

Here I have written down a few stories as an account of the actual events in which I participated. Some will find these tales to be at least entertaining, while others may learn something useful from them. But whatever your interest is, business or pleasure, I promise that reading these stories won't be boring.

In this book, I have only changed a few of the names, for I have not wanted to irritate the feelings of persons whose opinion and values might be different from mine.

This is my first nontechnical book written in English, which is my second language. It took me several months to write it, but many more to polish. This could not have been done successfully without the help of my editor Ralph C. Burr, Jr.

The thirteen chapters of this book are connected quite loosely, though they are written in a somewhat chronological order. Read them any way you please—forward, backward, or diagonally. I had fun writing this book and hope that the reader won't find that its first and last pages are too far apart.

Author

1
Making Movies

When I was a boy, my one and only dream was to become a film director. Actually, it was more than a dream—it was a passion. I was well aware that you had to know a great many things to make movies. So I would spend my free time finding out everything I could about motion pictures. I read about the history of photography and cinematography. I studied composition, lighting, editing, set construction, makeup, costume design—everything. I wrote my own screenplays. I devoured every book I could find about Chaplin, Eisenstein, Protazanov, Wells, and other great filmmakers. I found out endless fascinating things about motion picture technology: how movie cameras and film projectors were designed, and how to achieve various special effects. I even learned about the chemistry of black-and-white and color film developing, and about sound recording, mixing, and dubbing. Many years later, when I became an inventor, this early exposure to fine mechanics, optics, and chemistry turned out to be quite beneficial and helped me enormously. But at the time I was thinking only of the movies.

I studied acting and participated in a number of amateur productions at various children and adult clubs. I knew the plays of Shakespeare, Chekhov, Ibsen, and Ostrovsky by heart. I studied the acting system of the great theatrical director Stanislavsky. Some people were fairly impressed by my acting, and the school principal even engaged a professional actor from a repertory company to give me private lessons. This actor, Mikhail Zvantsev, taught me how to use my voice, how to move on a stage and in front of a camera, and various other tricks of the acting trade.

This knowledge came in handy on occasion. I remember one incident in particular. A very interesting film was being shown at the neighborhood movie house. It was a Czechoslovakian-made picture based on a book by Jaroslav Gasek, about the adventures of a brave soldier named Sweik during World War I. It was my favorite book, and missing the movie was out of the question. I had to see it by all means! The problem was that the movie was rated for audiences sixteen and older, and I was only thirteen. Ticket collectors in Soviet theaters were quite efficient about enforcing the rating regulations. I managed to buy a ticket, but getting through the entrance to the theater was another matter. Twice I was turned away by the vigilant ticket taker.

So I decided to age myself. I had a good collection of wigs, beards, mustaches, nose moldings, make-up paints, and other gadgets needed for changing an actor's appearance. I decided to make myself look just a little bit older—four, five years, tops. I did, I thought, a pretty good job. But though my face seemed older to me, the ruse failed. The ticket collector recognized me at once and said that if I dared to show up again, she would call the police.

I had no choice but to age myself more dramatically. I went back home and started transforming my face into that of a really old man. I put on a beard and mustache, added wrinkles, and so on. I knew how to create movie-quality makeup, which is a much finer art than that for the theatrical stage. Film makeup must withstand the scrutiny of a close-up shot, whereas theatrical makeup is only good from a distance. After about a couple of hours I looked like an eighty-year-old man. There was only one other problem—I needed clothes to match my face. And I did not want to apply makeup to my hands. That would be too complicated. My father's clothes were too big for me and my own were inappropriate. The only garment I could find in the house was an old padded winter jacket. Then I found felt snow boots, warm mittens, and a Russian fur cap. It was already May and quite warm outside, but what the heck—I had to see the movie!

At about four o'clock that afternoon, passersby were turning their heads at the sight of a funny-looking old man in winter clothes slowly shuffling down the street toward the movie theater. I almost jumped for joy when the ticket collector let me in. The film was great, and very funny. I chuckled, snickered, sniggered, giggled, tittered, cackled, and at the end started laughing uproariously. The man sitting next to me glanced over from time to time at this crazy old coot who was shrieking and laughing like a child.

At some point I noticed that my mustache had fallen off. The adhesive was not meant to withstand so much laughter. It was quite inconvenient. I still had to get back home and did not want to be caught by the ticket collector after such a successful deception! When the movie was over, I started coughing so that I could keep my mittened hand over my mouth, holding the mustache in place.

Drowning in my own sweat, I shuffled homeward, coughing and sneezing. Suddenly, I saw my grandmother coming down the street, on her way back from her shopping. At first she did not recognize this wheezing old man. But she knew the clothes immediately, and pounced on me. As she did so, she realized who the old-timer was, and grabbing my hand, she began dragging me home. People on the street must have been amused to see this sixty-year-old woman violently towing along a man who could have been her father. Back home we both shared a good laugh once I explained to her the purpose of my masquerade. She was quite an adventurous lady herself and could appreciate my persistence.

Besides playing such games in real life, I was playing various character roles on the stages of local semi-professional theaters. My burning desire, however, was still to make movies. At only thirteen I already knew most of the ins and outs of filmmaking, but there was no way I could actually make one myself. Movie cameras were not available for sale to private citizens in the Soviet Union in those days, and even if they were, my parents would not have been able to afford one. The only thing you could buy at that time was 16-mm black-and-white film. So I set out to design and build my own movie camera and film developing equipment. After hours spent in the library digging through technical books and learning how to calculate gears, springs, lenses, and so on, I came up with a set of drawings. I then asked my father to help me to turn my drawings into an actual machine. My father was a mechanical engineer in a large factory and knew some good machinists there. They moonlighted for him now and then, and bit by bit my dad started bringing me home the parts for my movie camera. I made a lens from a magnifying glass and used a crankshaft from an old Singer sewing machine which my grandmother donated to me. But the most impressive part was the case: a carpenter had made it out of a polished oak.

I designed a reversible camera. That is, I was able to shoot movies with it and also, after a simple conversion, to turn it into a movie projector. When the camera was assembled and tested, I bought fifteen meters of film, loaded it in under a blanket, set the camera on a tripod and raced outside. My camera looked very much like the one built by the Lumière brothers in Paris some sixty years before. I had seen a picture of it in the encyclopedia and was very proud that my creation resembled it. The first movie ever made had been of a train approaching the station. Similarly, I shot my first movie of a streetcar picking up its passengers. To work the camera, I had to crank the handle at a steady rate of speed, while at the same time keeping an eye on the focus. I then went home, mixed the chemicals, wound the film on a plastic frame and developed my first movie. I did all the cutting and editing, and then, using an electric lamp, converted my camera into a film projector. I enjoyed my first motion picture as I had never enjoyed any movie before, or probably since.

I had made just a couple of short films with my camera when the first commercial 16-mm camera, the "Kiev-16-S2," became available, and various clubs around the city began opening amateur film studios. Of course, the commercial camera with its spring wind and speed control was far superior to my home-made dinosaur, so I joined one of these clubs to gain access to this new equipment. Later on, I would also join the studio of the officers' club run by the district military command of the Soviet army—but more about my memorable adventure there in the next chapter.

Quickly mastering the 16-mm camera, I made a number of short documentary and dramatic films. In the winter of 1961, there was a total solar eclipse in our area. Everyone was looking forward to seeing this rare event. I was excited as well, hoping very much to get the eclipse on film. Obviously, the optical power of a 16-mm camera was not nearly strong enough for filming astronomical objects, so I assembled a kind of telescope with a black filter. On the frosty day of the eclipse, I was in the school yard with my camera pointed toward the cloudy sky. I was not the only one hoping to photograph the eclipse. The local TV station was also attempting to capture the fading sun on film to show on the evening news. But all in vain—the clouds were too thick and our optical equipment inadequate. No one there had any luck. The image of the sun on my film came out too small and very hazy. The TV people fared no better.

I was disappointed but not discouraged. Once a month I showed my ten-minute "movie magazine" to the entire high school. My friends had seen me shooting the eclipse, and I did not want to let them down. Resorting to my knowledge of special effects, I assembled an apparatus at home using a sheet of cardboard with a round hole in the center, a piece of cellophane, an electric lamp, and a couple of smoking cigarettes to imitate clouds, and in just one evening produced my own "solar eclipse." I developed the film the same night and it came out great. You could see clouds, and the big round disc of the sun, and the shadow of the moon passing across it—and even a solar corona! It was so realistic, in fact, that I decided to take it down to the local TV station. The chief producer for the news program was very impressed with my success. Of course, I never told him that it was just trick photography. That very night they put my "solar eclipse" on the news. The studio's entire camera crew was mortified—even with all the finest equipment of the time they could not get a single decent shot of the eclipse, and here a sixteen-year-old schoolboy had captured it beautifully.

Some will argue that it goes against all journalistic ethics to fake the news. I agree. In my defense I can only say that in a country where most of the news was not only planned but often fabricated, faking a solar eclipse seemed fairly innocent. Especially if one was not a journalist and only sixteen. So I cannot feel too badly about it. Many years later my sister-in-law's father, who had worked at the TV station as a sound director, was visiting us in America. I was amused when he told me that even now, thirty years later the old-timers at the station remember those incredible shots. He said the cameramen had always suspected some kind of trick but could never figure it out.

Before long a TV news producer offered me a job as a freelance correspondent. Thrilled, I began my journalistic carrier while still in the ninth grade. I worked at the station for about six years. The usual evening news program consisted of several two-minute segments. In those days, all news events were filmed with 16-mm cameras. These cameras had no audio capabilities, so all the footage was like something out of an old black-and-white silent film. My responsibilities were quite demanding: I had to find an interesting subject, arrange the lighting, shoot whatever would be needed, interview the subjects, write the text for the news anchor, edit the segment, and get it all ready for the censors by five o'clock on the afternoon of the broadcast. Eventually, I was working as correspondent and cameraman at the same time. Sometimes the studio lent me a professional 16-mm camera, developed the exposed film, selected music and noise effects for the looping, and when needed, provided the help of a lighting assistant. They paid me ten rubles for each segment. At a time when the average Soviet worker was making about 120 rubles a month, this was pretty good money for a teenager. Needless to say, I was quite productive, and my newsreels were broadcast every week.

Soviet journalistic methods were very different from those of the West. Correspondents never broke their necks to get actual news. Normally, a producer would distribute the daily assignments to the staff correspondents, who then had to go out and deliver whatever was needed by evening, without showing any real initiative. In the Soviet Union, all news had to be good news, and all good news was planned ahead. All bad news was supposed to come from the capitalistic countries, and it was the function of the Central TV Station in Moscow to supply the local stations with the appropriate materials. Our job was to produce only good news of local interest. But I enjoyed filming so much that such restrictions were immaterial to me.

Usually the editor would call me at home, or else I would stop by his office after school to get my assignment, if he had one for me. As a rule, I got whatever was left after the staff correspondents had been given their tasks. Naturally, they always got the easier and more pleasant jobs, but that was fine with me. Often I would go actively looking myself for something I could use as a subject. Of course, whatever I brought back to the studio had to pass by the censors as well as conform to the standard of good and optimistic news. But even within such a restrictive framework I managed to find a great variety of interesting subjects.

In 1963 I was working for a large factory while at the same time studying as a sophomore at the Polytechnic Institute (the equivalent of a university). In addition, I often moonlighted for the TV station. In May of that year, it was announced that the Cuban president Fidel Castro was to visit our city. The Cuban revolution was a central theme of Soviet propaganda in the early sixties. In the wake of the exposure of Stalin's crimes, many people had become disillusioned with the ideas of Communism. There was a feeling that the only way Communism could be established and defended in other countries was with Soviet tanks, as had happened in Hungary in 1956. Castro, with his inclination toward Marxism, was a blessing from the sky for the Soviet Communist Party. At last they could say, "See, it has nothing to do with our tanks. It's the will of real people many thousands of kilometers away from the Soviet Union." At the same time, Nikita Khrushchev was trying to butter the Cubans up, because Castro was very disappointed when the Russians gave up and pulled out after the Cuban missile crisis of 1962. So Castro was given the full royal treatment during his state visit of 1963. Indeed, it was more than royal—he was treated like the greatest man who had ever lived. The entire Soviet Union went wild when he and his bearded comrades arrived.

When I asked at the TV station if there was any way I could go and report on the visit of the Cubans, the editor just laughed. "First of all," he said, "if we send someone, it will be a staff reporter, not you. Secondly, we are not allowed to get anywhere near the Cubans. This is an international event and the only ones covering it will be Moscow Central TV."

"Aha," I said to myself. "The studio's out. But I have to be there. I want to see for myself what kind of people these Cuban heroes from the Island of Freedom are." But how was I to get close to the delegation, which was surrounded by a double ring of security, both Russian and Cuban? Then I remembered a friend of mine from high school, Krotov, who had a big-shot father. Later minister of heavy industry, Victor Krotov was at the time a managing director of Uralmash, an enormous factory which produced various heavy machinery from tractors to tanks. The Cubans were supposed to visit the factory on May 23—so why not to talk to Krotov?

I called my friend and asked him whether his father could arrange for me to shoot the Cubans? (Of course, I didn't mean *shoot* them, just film them, otherwise I would never be here telling this story now.) He said, "I'll talk to my dad. Call you in a half hour."

A little later he called back and gave me the good news. "Dad says the delegation arrives at 9 A.M. tomorrow. You must be at his office at 6 A.M. He will tell the guards to let you in. But don't be late."

I called the TV station and asked the news producer to lend me a 16-mm camera. But when he heard what I wanted it for, he flatly refused. "I don't want to hear this. We are not allowed to get close to them. If you want to do it, you are on your own and we don't know you." That was that. I could get close, but I had no camera.

At the time, I was a member of the film studio run by the university students' club. We had a few old cameras, including an "ancient" French-made Debrie Parvo. Of course, none of the equipment was as good as the modern cameras the TV station used. When I went to the university studio the only available camera was an old 35-mm Konvas-Automate. Actually, the Konvas was a terrific camera, with three lenses on a turret, a reflective shutter, and an electric motor drive. But our Konvas had seen better days. It had been banged about a lot, and was rusty and awfully noisy. Besides, the only battery we had was a huge lead-acid box weighing about forty pounds. But I had no choice. I grabbed the camera, the battery, and several cassettes of high-speed film.

The morning of May 23, I arrived at Uralmash right on time. The guards escorted me to Krotov's spacious office and told me to wait there. When Krotov himself arrived at eight o'clock, he seemed very surprised to see me. "Who are you?" he asked.

I reminded him that I was there at his own invitation. Remembering, he said, "Oh, yes, yes. That's right. But stay out of trouble, and don't leave this office. If security catches you without any papers I wouldn't want to be in your shoes. The Cubans will come here after touring the factory. Actually, they'll be spending some time in the conference room next door. You'll have a good opportunity to film them then. Good luck."

I waited for another hour. Adjoining the office was a large conference hall. A long mahogany table set with fresh flowers and bottles of mineral water ran down the center of the room. The tops of the bottles were wrapped in foil, just like Champagne. A group of security men came in and began checking every corner, every chair, every bottle, and all the windows. One of them was carrying a metal detector and a radiometer to detect the level of radiation. It seemed the Cubans were concerned about visiting the Urals, where Soviet nuclear weapons were developed and produced, and where the radiation level was higher than in other parts of the country. Since I was already in the office, the security men never asked me for my papers, assuming I was supposed to be there, but they scanned my body head to toe with their instruments. They were especially curious about my huge battery, but everything checked out as normal. The conference hall had French doors opening onto a balcony, and from there I watched as a long procession of black limos brought the visitors straight to the factory.

After another hour, the large doors to the conference room opened wide and more security people came in. Then I saw a gray-haired man whose appearance excited me more than anyone else. It was Roman Karmen, the famous cinematographer, who had made a number of great documentary films. I had seen many of them and had read his books, and knew that he was a friend of Castro's. Karmen was known for shooting under the most dangerous circumstances. During World War II he captured the battle for Berlin and the capitulation of Germany on film. He had filmed in Vietnam and Korea and in the Cuban mountains when Castro was not yet a powerful dictator, but just the guerrilla Fidel. Karmen was wearing in light suit and had a Konvas-Automate slung over his shoulder. My God, what a beauty that camera was! The Konvas was all white, with glittering chrome parts. The Nicad battery on his shoulder was light and sleek. Seeing me with my black and rusty old camera, Karmen could not hold back an ironic smile. His two assistants quickly set up bright floodlamps that lit up the whole room. I was getting a free ride with that illumination. No other cameramen or correspondents were being allowed into the building. I stepped out onto the balcony and saw some fifty journalists, both Russian and foreign, down there on the pavement. I was so proud when they saw me standing next to great Roman Karmen! They were dying of envy! Some even called to me to take few shots for them.

A few minutes later, the big doors opened again, and Karmen and I put our cameras on our shoulders and got ready to start filming. Castro and Krotov walked into the conference hall at the head of a large crowd of Cuban and Soviet officials.

What happened next was one of the most bizarre events of my life. As soon as I pushed the "motor" button on my camera, that damn rusty Konvas suddenly let out a terrifying and ear-piercing roar, just like a machine-gun: "Tra-ta-ta-ta-ta-ta-ta!" In the same instant Castro fell to the floor. Two security men pounced on me, twisting my arms and pulling the thundering Konvas out of my hands. I was lucky they did not shoot me on the spot!

However, this dramatic and unexpected accident was quickly and peacefully resolved. The Cubans lifted the unconscious Castro and signaled to the security men to let me go. They explained that such things happened to Castro, that he often fainted when it got too hot, or when he was particularly tired, and that it had nothing to do with me. It was sheer coincidence that my "machine-gunning" and his fainting spell had happened at the same moment. The Cubans lay Castro on the couch, removed his boots and poured cold mineral water all over his bare feet, straight from the foil-covered bottles. Everyone breathed a sigh of relief. The Cuban prime minister, Osvaldo Dorticos, came over to me with a sort of apologetic smile and, through Nikolai, the interpreter, asked me not to use my strange camera any more, at least inside. I thought Roman Karmen was going to die laughing. Not wanting any more trouble, I quietly set my Konvas in a corner of the room and just hung around till the end of the meeting listening to the small talk between Castro, Krotov and Kuznetsov, the Soviet deputy foreign minister.

When barefoot Castro had rested, he asked me, "I hope I didn't scare you too much?"

"It looks like I scared you even more with my toy," I said.

He laughed, holding his cigar between yellowish teeth.

After an hour or so, we all went outside to the large square for a public meeting. I started shooting again, but it was not of great interest and no one was scared anymore by my Konvas.

Needless to say, I became an instant celebrity at the university and the TV station. Not only had I gotten to talk to the legendary Castro, I had even seen him lying unconscious! Who else had that kind of luck? When the negative was developed, it was far too brief to make any meaningful news segment. I ended up simply cutting a single frame from it and printing a photograph of Castro lying on the floor.

Since I was a student at the Polytechnic Institute, the news producers saw me as a natural choice to cover various technology-related events. They were unperturbed by that fact that a 19-year-old student might not be terribly knowledgeable and that his reports had better be checked over before broadcasting. One day this negligence landed us all in a good deal of trouble.

The chief news editor, Leon Kogan, asked me if I could find out anything interesting at the university about new developments in science and technology. I said sure, grabbed a camera and took off. I headed straight for the university design center, where the undergraduate and graduate students worked on various high tech projects, many of which were subcontracted by industry. One big project was the development of a camcorder; another, a music synthesizer.

The center, located in the basement of the radio-electronic department, was bustling. I knew the guy who was the technical manager of the center and asked him if he could suggest something for a news story. Scratching his head, he said that everything they had was too ordinary and not particularly newsworthy. Suddenly, he slapped his forehead. "Talk to that red-haired kid in the corner of the room. He may be what you need." I went over to a student who was sitting at a bench buried in a tangle of wires and magnetic tape.

"What are you up to here?" I asked. "Is it anything exciting that we could show on TV?"

"You bet," said the freckle-faced nerd. "I'm making a typewriter with a microphone and a small electronic box." He pointed to a funny-looking typewriter without a keyboard.

"You see," he said, "I don't need a keyboard. You just talk into the microphone, and the electronic box will translate your words into the appropriate key strokes and the typewriter will print the text."

"Can you show me how it works?" I asked.

"Not yet. I need a few more months to work out some minor problems."

Well, I did not realize then the magnitude of those "minor" problems. Computers capable of speech-pattern recognition did not yet exist at that time. But I was naive enough to believe the student. I phoned the TV studio, they sent me a lighting assistant, and I did a report about the red-haired genius and his keyless typewriter.

My segment turned out quite well. In fact, it was so impressive that the station not only showed it on the evening news, but even called Moscow Central TV and suggested they run my report nationally and perhaps internationally. I was thrilled. Apart from everything else, my fee would be doubled. And they did it. Moscow ran my report about the typewriter throughout the entire Soviet Union.

Three days later, my mother met me when I came home from school and told me to hurry over to the TV station right away. They were looking for me, and whoever had called had been very nervous. When I got to the station everyone turned to me with a strange look, and Leon, the editor, hissed me, "Follow me, you little S.O.B.!"

We went up to the second floor, where the plush office of the chairman of radio and TV broadcasting was located. The secretary ushered us straight inside and there I saw the big boss along with the pale-faced head of the censorship section and two middle-aged gentlemen wearing civilian clothes and bulletproof faces. The chairman introduced one of them as a KGB general responsible for the city and the region. The other was a colonel from Moscow, apparently from the same illustrious organization.

The colonel addressed me, "Young man, how do you explain the fact that a technological development of the utmost state importance has been publicly disclosed? We have several research institutes working full time on speech recognition, and here you go telling the whole world about a student who's about to solve the entire problem! This project is supposed be classified as top secret, and you and your stupid news story have gone and made the whole thing public!"

These security people clearly had little appreciation for the complexity of technology. Otherwise, they would never have been so quick to believe that such a fundamental problem could be solved by an undergraduate student. Even one with freckles. I tried explaining to them that we had simply been duped, that kid had been pulling our leg and we had fallen for it. But the visitors had little interested in what I thought, especially since I was a freelance reporter. They sent me out of the room, and spent the next fifteen minutes or so lecturing the station people on how to handle information about new technical developments. Since there was really nothing to the whole incident, the KGB never pursued it any further, and no one at the station was punished. Still, the entire news crew was so furious at me that it was some six months before I dared showed my face there again.

During all those years freelancing for the TV studio, I never had much luck with the censors. Whenever I got a really interesting subject, or was just able to do something out of the ordinary, the censor was there with his scissors and big rubber stamp. In 1966, the student club suggested that I go to Hungary to make a documentary film about the country. The TV station said they would gladly broadcast my film as well when it was ready. The prospect of such a trip was tremendously exciting, both for me and for my parents. The mere fact of going abroad, even to a socialist country, was a rare and unbelievably lucky event for a Soviet citizen. However, the red tape was endless. It took me nearly three months to get all the required approvals. Finally, in July, I left for Budapest. Hungary is a beautiful country with a rich history, and I was kept very busy shooting during my nearly two weeks there.

Everywhere I went in Budapest, I saw green flags and intriguing posters displaying the logo of the World Esperanto Congress. I had heard something about this artificially created language, but knew little about it. One evening while I was having my dinner in a cafeteria, I noticed a large group of men and women in green outfits eating at a nearby table. It turned out they were the Bulgarian delegates to the Congress. They were a friendly group, and we struck up a conversation. They told me about the optician Ludwik Zamenhof from Poland who invented Esperanto in 1887, and explained that the language was very simple and that it was based on several European languages. It was all very interesting, and I was captivated by the beauty of the language structure and the melodiousness of its sound. The Bulgarians told me that the language was so easy that Leo Tolstoy had learned it just in a week, and that an ordinary person could get it in a month or so.[1]

I returned home determined to learn Esperanto myself. I went to the public library, and there, amid the stacks of a storage room, I found several old and dusty Esperanto textbooks published in the early twenties. It was indeed a simple language, and after about a month I was already reading it without a dictionary. But I also wanted have a conversation with someone, yet I could find no one in our city of a million inhabitants who knew the language. I could speak it, but only to myself.

One day, somebody told me about an old man living at the outskirts of the city who spoke the language. I decided to go see

---

[1] Ironically, Zamenhof, a Jew, invented his artificial international language in the hope that it would help to unite the world and that everyone would speak it. Another Jew, Ben Yehuda, pioneered the use of Hebrew in Palestine. Hebrew was a language few people spoke at the time, except during religious services. Today, Esperanto is spoken by almost no one, and in Israel, meanwhile, practically everyone must know English, for Hebrew has no significant use either in international relations, or in literature or science. It would appear that languages, no matter how beautiful or ancient, cannot be artificially imposed.

him. The man, who was in his late seventies, lived rather poorly in a small room in a communal apartment, sharing a kitchen and bathroom with four or five other families. When he escorted me to his room, I started talking to him in Esperanto. Looking shocked at first, he peered into my eyes, then, without saying a word, he grabbed a tablecloth and thumbtacks and covered the window tightly. "You must be careful," he said in Esperanto. He told me that the language had been very popular at the beginning of the century, but that in the thirties Stalin accused anyone who spoke Esperanto of being a spy. Virtually everyone who knew the language was exterminated. The old man in front of me was one of the few survivors. He had spent fifteen years in a concentration camp beyond the Arctic Circle, and spent the rest of his life living in fear—thus the tablecloth over the window. All the same, I was overjoyed at last to find someone who understood what I was saying in this beautiful language.

Naturally, I wanted to do a TV show about Esperanto and tell everyone all the wonderful things I had learned about it. The studio producer thought this was a good idea, and for two weeks we worked on preparing the show. We decided that I would go on the air live with part of my film from Budapest. Then I would present various books in Esperanto and even read some poetry in the language.

Everything was soon ready, and on the day of the broadcast I was sitting under the bright lights in front of the TV cameras (in those days most shows were broadcast live, without the use of video recorders). Just a couple of minutes before we were to start, we saw a short plump woman rush into the studio with a huge book in her hands. She was one the station censors, though actually she reported directly to the KGB. She shouted, "Stop! Stop it! I forbid this show to go on the air!" She pointed at paragraph number so-and-so, which stated that any mention of Esperanto, either in print or in a broadcast, was strictly prohibited. The date of the paragraph was 1937.

The show's producer and I were flabbergasted: "What are you talking about? This was nearly thirty years ago, during Stalin's Great Purge. Those practices were condemned long ago."

"I don't care," she said. "I go by the book. Nobody ever rescinded this paragraph, and there's nothing I can do about it."

That, then, was the demise of my Esperanto show.

When I was about to graduate from high school, I decided to apply to the Moscow film school, the only such educational institution in the entire country. It was and still is known by the abbreviation "VGIK," which stands for the Russian for "All-Union State Institute of Cinematography." A VGIK education was an essential step in becoming a professional film maker. Without a diploma from this institute I could never go on to be a director. Getting into VGIK, however, was extremely difficult. The school's admissions process was corrupt. Anyone without connections had to be thoroughly exceptional to get in. But I was young and still had an optimistic view of the world—which in this case meant an unrealistic view. So I decided to apply anyway, and prepared the detailed screenplay and shooting plan which one had to submit along with the application. The VGIK held a competition of these works, and only the winners were then allowed to take the entrance examinations.

My mother was very worried about my passion. Time and again she said to me, "The movie business is too risky. It's okay to be an average engineer, but you can't be an average movie director. There's no place for mediocrity in art. How can you be sure that you are so superior?"

I listened to her words, but still could think of nothing else. She decided to go to Moscow to visit the VGIK and see for herself what the place was like and what kind of people taught there. She even managed to talk to Mikhail Romm, a famous director and a professor at the institute. I knew all of Romm's movies. He explained to my mother frankly what was involved in becoming a student and what difficulties I might face trying to get in.

The year was 1962, and suddenly Nikita Khrushchev, with his zeal for crazy, nationwide innovations, made my decision for me. He enacted a law which stated that no high school graduate could enroll in any university or college without first working two years in industry. This automatically precluded me from applying to VGIK. I had no choice but to postpone my attempt by at least two years.

My parents convinced me that it would be good, in any event, to have a back-up profession. If, God forbid, my film career failed to pan out, I would still have another, "normal" profession in which to make a living. Besides motion pictures, I had something of an interest in medicine and biology, so medical school was one option. However, an M.D. as a back-up profession did not seem terribly wise. To me, medicine involved virtually the same moral responsibility as art. You had either to be very good, or forget it. But if a film director is less than good, only he suffers. A poor doctor, on the other hand, can make many people suffer—and this was unacceptable to me. If I chose medical school, I would have no choice but to become the best doctor I could. And that meant forgetting about film school.

Medicine, therefore, did not fit my definition of a "back-up profession." Instead, I decided to enroll in the Polytechnic Institute, which was the equivalent of a technical college. The Ural Polytechnic Institute in Sverdlovsk was a fairly good engineering school, with some twenty-five thousand students. It had several departments, among which physics and radio-electronics were the strongest. But most important, it had evening classes in all departments. Khrushchev's new law allowed one to take evening classes during the two years one had to spend working in industry. Thanks to that loophole, I could both be a student and serve my industry time simultaneously, which meant that at the end of the two years I could start my efforts to get into the VGIK.

So that's what I did. I applied to the radio-electronics department at the Polytechnic Institute, passed the entrance exams, and was admitted. At the same time, I got a job as an electrician at a huge factory that made electric motors and generators for power plants.

My workday began at 7:30 AM. After working my eight hours I took a streetcar home, rested briefly, did my homework, then headed off to school at 7:00 P.M. It was an intense schedule, considering that I was also still working as a freelance TV news correspondent, as well as performing in semi-professional theater productions and, in addition, becoming very active in the student film studio.

The studio belonged to the students' club at the institute. It had several used 16- and 35-mm cameras, floodlights, a moviola, and some old sound recording equipment. We made documentary, educational and sometimes dramatic films. Before long, I had switched almost entirely to the 35-mm format, reserving the 16-mm for TV reporting only. We usually shot all the scenes for both documentary and dramatic films without sound, then later would rent facilities and equipment from the commercial movie studio and do all the looping, editing, special effects, and so on there. One of my dramatic films, "The Impudent Experiment," even received first prize at a festival of short dramatic films.

Shooting "The Impudent Experiment"

When the two years had elapsed, I quit my job at the factory and became a day student at the Polytechnic Institute. At the same time, I was working very hard on the screenplay and director's shooting plan that I had to present with my application for admission to film school. I enjoyed studying engineering sciences, but motion pictures were still my passion. The screenplay I wrote was based on Andrey Bitov's short story "In the Train." It was a tale of love at first sight.

When I had everything ready, I gave the manuscript to my friend G. for his comments. This G. was working on his own screenplay, for the same purpose. He was four years older than I, and was making his fourth attempt to get into the VGIK. Always very secretive, he never told me what his screenplay was about or how far along he was. He was a friendly fellow, but rather lazy. I had often seen him drunk. Nevertheless, he and I shared the same goal, and in that sense we were in the same boat. I regarded him as a kind of traveling companion and wanted to run my screenplay by him.

He held onto my materials for several days, made a few interesting suggestions, and then a couple of weeks later, after some more polishing, I sent a big brown envelope containing the fruit of my many months of hard work off to Moscow. All I could do now was wait for a letter from the VGIK telling me whether or not I had been allowed to come take the entrance examinations.

Two or three months went by, and to my great surprise my friend G. one day received permission to take the examination. Apparently, his screenplay and shooting plan had won the very intensive competition! It was an incredible stoke of luck. It by no means guaranteed his admission, but it was the most crucial step toward getting into the school. I received nothing, however, and began to be worried. I wrote a letter to the school asking about my fate, but again there was no reply.

Before long, I noticed a change in the way people at the Polytechnic Institute behaved toward me. Some avoided looking me in the eye, others refused even to talk to me. I felt like the carrier of some shameful and contagious disease, but could not begin to guess what was going on. It was a complete mystery—until one day I cornered the president of the students' club and asked him point-blank what was going on. He looked at me with a strange smile.

"Don't pretend you don't know," he said.

"Know what?!" I became hysterical. "I don't know a thing, and you're not going anywhere until you tell me everything!"

He then told me about a letter which had been sent to the rector of the institute about a month before by the rector of the VGIK informing him that the film school had received a screenplay and director's plan entitled "In the Train" from the student G. This G. had also included a letter in which he warned the examination commission that a certain student by the name of Fraden had secretly copied his materials and was planning to send the same thing to the VGIK. Sure enough, a couple of weeks later they received a brown envelope from Fraden with an exact copy of G.'s screenplay. The rector of the film school wanted to bring my "disgusting and shameless" behavior to the attention of our rector. The letter also said that I would never be considered as a VGIK student. The contents of that letter had apparently become known to everyone, except me.

What could I say? It was like being shot through the heart. I thought I was going to die. For a long time I stood there in the corridor, deaf, dumb, devastated. Then I ran through the entire building looking for G. I found him in the cafeteria. Grabbing him by the shirt, I yelled, "You stole my scenario, you scumbag!!"

He looked me straight in the eye without blinking and said quietly but firmly, "No, you stole mine, and you know it." With that, he turned and walked away.

I went to see the rector and asked him about the letter. He told me that he had indeed received such a letter, but that he had felt there was no way to prove it anything one way or another, so he had just filed it away without telling anyone.

"But why does everyone knows?" I shouted.

"I have no idea," said the rector.

I later found out, of course, that it was G. himself who had spread the rumor and organized "public opinion" against me. The incident marked that loss of my moral innocence. I was crushed and had no idea what to do. There was no way I could prove anything or convince anyone. Indeed, no one seemed terribly interested in knowing the truth. I tried talking to people I knew, telling them what had happened. Some believed me, some did not, while others simply shrugged and said they did not care.

Time passed, and everyone but me forgot about this minor incident. People have short memories, and few, indeed, really do care.

A few days after confrontation with G. in the cafeteria, he flew to Moscow, where he successfully passed the exams and was admitted to the VGIK with my screenplay. I cannot say what happened to him after that. I have never seen any movie directed by him, nor noticed his name listed in any film titles. I have no idea whether or not he graduated, or whether he ever made any movies, or even whether he is still alive. Nor do I care anymore.

But I am positive of one thing. His treachery was a blessing for me. In my mind, my entire world had been destroyed, but soon a new world began opening up to me. It was like "The King is dead. Long live the King!" Pulling myself together, I went on studying at the institute, went on making my films and freelancing for the TV station, went on performing on stage. I continued doing what I loved to do—but now my back-up profession began moving into the foreground. I decided to become an engineer. Often in the course of filming my TV news segments I would visit research laboratories, where I met interesting people and saw their work. I began to understand that there was room for creativity in engineering. Designing machines or inventing new devices could be just as exciting as making movies! And electronics, which I was studying, was the best of all. Its toolbox was enormous, and its possibilities limitless. Making movies, I discovered, was just one of many ways of being creative. Doing something original and at the same time useful was what was most important for me.

And what about my dream? I put it aside indefinitely. Maybe someday...

Many years later, in America, I met M. B., a famous Russian film director. He had graduated from Mikhail Romm's class at the VGIK the same year I graduated from high school. After making several internationally acclaimed movies, he had become very frustrated with the ever-tightening control of the arts by the Communist Party. His creativity virtually stifled, in the late seventies he decided to emigrate.

The Soviet motion picture industry virtually collapsed along with the Soviet Union. Today, the art of Russian film is a shambles. Those cinematographers who stayed have ended up in ruin and disarray, while those who emigrated have had to change professions in order to survive. Even in America, the life of movie makers is not very easy. The unemployment rate among actors and directors is well over ninety percent. Even the very good ones need incredible luck. But luck comes so rarely. M.B. came to this country with the hope of finding a place for himself in Hollywood. But who needed a Russian cinematographer in America?

M. B.'s life in the Soviet Union had become impossible, but it turned out to be very hard in America as well. For a time he worked odd jobs as a restaurant dishwasher or as a night guard at a warehouse in the South Bronx, and he considered himself very fortunate whenever he could get some small gig at a Hollywood studio. I looked into his beautiful but sad and tired eyes, and thought to myself, "The same thing could have happened to me if I had succeeded in my dream." Had I not been deceived by G., had I graduated from the VGIK and become a film maker, where would I be now? We have no way of knowing what is good for us. Today, looking back, I feel that there must be Someone up there watching over me and protecting me. Life takes some funny twists and turns—and sometimes our deepest miseries become our greatest blessings.

2

The First American

The officers' club was only a mile from the apartment building where I lived with my parents and sister. I used to pass by it every day on my way to and from school. One afternoon as I was walking home a sign on the notice board near the main entrance to the club caught my eye. It announced that the club was opening a film studio and anyone interested in making films was invited to join. I was very excited by the prospect but at the same time afraid that an officers' club would never allow a schoolboy into its studio. Nevertheless, I decided to give it a try.

The officers' club, or, as it was officially called, the House of Officers, was located in a magnificent, even posh building in the center of the city of Sverdlovsk, now Yekaterinburg. The architect's design was fairly ingenious. From above, the building and the nearby military residential complex formed the shape a hammer and sickle—the logo of Communism—but this was only visible from a plane or helicopter. The granite outer walls of the building were rich in architectural detail, while inside the club was filled with priceless furnishings, crystal chandeliers, marble staircases, redwood handrails—you name it. The club belonged to the Staff Headquarters of the Ural Military District, which had its offices just half a mile away.

Despite my fears, the guard at the entrance looked me over without surprise. Family members of the District Staff officers were frequent visitors at the club, and a teenager drew little attention. The guard directed me to a marble staircase and told me that the studio was on the third floor and was now open. I went upstairs, knocked on a mahogany door with a fresh sign that read "Film Studio," and heard someone answer, "You may!".

Inside the small room, I found a portly bald man about sixty years of age, dressed in military-style breeches and a Ukrainian shirt. He was sitting in an armchair, smoking a cigarette. A "No smoking" sign hung on the wall above his head, with a poster below it depicting a grinning pig with a cigarette in its mouth. The pig was saying, "But I smoke anyway!" The pig in the poster and the bald gentleman in the armchair bore a curious resemblance to one another—a situation that seemed to indicate that the man possessed a rebellious sense of humor, not a common quality in a military person.

I introduced myself and said I was interested in joining the film studio.

The pig-like man gave me an ironic smile. "I am Lieutenant Colonel Grigory Pavlovich Boyko, commander of this studio. How old are you?"

I replied that I was not quite fifteen, but that I was no novice when it came to shooting movies, or, for that matter, anything related to the production or projection of films.

Boyko smiled broadly and barked, "We'll see about that. Here's a 35-mm projector. Can you load the film into it with yours eyes closed?"

I did it in less than thirty seconds. Boyko was impressed. "Well then," he said more gently, "can you load a cassette for a 16-mm camera?" He pointed to a Kiev-16-S2.

I knew this camera very well and quickly loaded the cassette in a film changing bag. When I was finished, Boyko told me I could join the studio on a trial basis, and that he would see how I was doing before deciding if I could become a permanent member. That I was underage and not in the service was no problem, since the studio was an open organization with no access to classified materials.

Before long, I had become an indispensable member of the studio, involved in shooting, editing, looping, and projecting virtually all the movies produced there. The main function of the studio was to make monthly newsreels covering events within the Ural Military District. These newsreels were shown in all the regiments, divisions, and other military sections of the District. The studio was very well equipped. Most of its production was on 35-mm film. The cameras and the editing and sound equipment were all brand new, and of the same quality and standards as those used by commercial film companies. The studio had only one paid staff member—Lt. Col. Boyko. Besides Boyko and me, there were two other men who shot military news subjects. They were officers stationed at the Staff headquarters, and whenever a shoot involved travel to military installations, field exercises, training missions, or the like, they and Boyko went without me. Being underage and not an officer, I only worked on the shoots dealing with nonmilitary subjects. However, I carried out most of the other functions related to the production of the newsreels. Technical services such as developing the film and cutting the negatives and printing the positives were contracted out to a commercial, civilian studio.

Lt. Col. Boyko and I had a good relationship. I became his right hand—he relied on me totally, and nothing ever happened in the studio without my participation. It was a situation that served both of us quite well. I was more than happy to work with the professional equipment producing movies for the big screen. Boyko, meanwhile, was lazy and always slightly drunk, so my eagerness was very convenient for him.

On the last day of April 1960, Boyko told me that he and I had to be at Lenin Square the next morning to film the May Day military parade. Soviet military parades were enormous propaganda shows with goose-stepping troops, tanks, heavy artillery guns, anti-aircraft missiles and other modern military equipment. All this machinery rolled across the stone-clad pavement, while the top Party and military bosses watched from the top of a granite podium. The whole thing was designed as a big exhibition, and we were supposed to get it on film for the next newsreel. The plan was that Boyko and I would meet at the studio early in the morning to pick up all the photographic equipment and load it into a van that would take us to the square. Once there, we had to film the parade, then take the equipment back to the studio. After the holiday, on May 3, we would take the film to be developed at the commercial movie studio, located just two blocks from the officers' club.

On the morning of May 1 I arrived at the studio shortly before seven as we had agreed. I waited some half an hour, but Boyko never showed up. I called his home. His wife answered the phone. She said the lieutenant colonel was still asleep and that she really did not feel like waking him. I explained to her how important it was, and that we had to be at the square soon. If we were late, all the roads would be blocked and there would be no way for us to get there in time for the filming. Finally, she agreed to wake him, but when the lieutenant colonel picked up the phone, I realized that his May Day celebration was already well underway.

"Ah?" he said. "Who is it? Wh-wh-at do you wh-want? Oh, yeeees, sh....., I am, a-a-am... I am sick, you see, Jake-kid? You gonna go by yourself, right? S-s-so, don't wh-wh-ait for me... I am sick, got it?" And he hung up.

He was drunk as a pig (do pigs drink?), and knowing him as I did, I was hardly surprised. Anyway, I did as he told me. I grabbed several cassettes with film for outdoor shooting, the case with the 35-mm camera, the tripod, and loaded it all into the van. The driver took me to Lenin Square. Along the way, we were stopped several times by soldiers checking my pass. During military parades, security around the square was usually very tight.

When we got to the square, I set up the tripod and mounted my 35-mm Konvas on it, then attached the cassette and prepared start filming. Several cameramen from the commercial movie studio were also there. In those days, television had no live broadcasts, at least in the Soviet Union, so apart from me, there were only a couple of film crews and a group of newspaper photographers at the square. When the parade started, I filmed the most colorful scenes: ballistic rockets with fat bellies borne along on huge trucks, and portly party leaders with equally fat bellies up on the granite podium.

After I was finished, I just hung around waiting until I could load my stuff back into the van and take it back to the studio.

Just then, I saw an officer with a captain's epaulettes running from one film crew to another, apparently looking for someone.

Finally, he came up to me and asked, "Are you from the House of Officers movie studio?"

"Yes, I am."

"Where is Lt. Col. Boyko?"

"He's... Well, he's sick at home."

The captain swore.

"Are you shooting here by yourself? Can you do indoor shooting? Right now."

I said yes, and he asked if I needed any additional equipment. I told him that the only thing I might need was some high-speed film and a couple of floodlights. He then informed me that I was to go with him to photograph a very important meeting and that whatever I required would be arranged. We quickly packed up my equipment which two solders helped me load into a military van with tinted windows, and we were soon speeding away from the square. Obviously the van must have had some special plate, because we drove practically nonstop for about five minutes.

When the van stopped and the captain let me out, I recognized the place at once. We were behind the Staff Headquarters of the Ural Military District, just a few blocks away from the House of Officers. The two soldiers helped me get my equipment out of the van and took it up to the second floor. I was led into a spacious room where the captain ordered me to get set up and wait for further instructions. He then left, and I began assembling the camera in one of the corners. The room had large windows covered with heavy red curtains, and a long mahogany conference table in the center with about a dozen chairs around it. A portrait of the Soviet Prime Minister Nikita Khrushchev hung on the wall. Some officers brought in a large tape recorder and placed it on the floor near my camera, attaching two microphones to it. They then placed the microphones in the center of the table.

I had been waiting for about twenty minutes when the door opened and Lt. Col. Boyko walked in with a large case. He was followed by two soldiers carrying floodlights. To my surprise, Boyko appeared quite sober, yet the greenish color of his face and a strong smell of vodka gave him away. His hands were trembling.

"Listen, kid," he said in a hoarse voice, "I brought you some high-speed film, but you see, I can't load it into the cassettes. So here's the bag, you do it. We may need a lot of film. And do it fast."

"What are we going to be shooting, and for how long?" I asked.

"I don't know. They just told me we have to get some kind of meeting on film, that's all I know."

In his case, I found six empty cassettes and two metal canisters of high-speed 35-mm negative film, sufficient for about twenty minutes of continuous shooting. It took me about fifteen minutes to load the cassettes, while Boyko and two soldiers set up the floodlights.

I had just finished loading the last cassette when the doors opened and several officers entered the room and motioned to me and to the men operating the tape recorder to start rolling. As Boyko turned on the lights and I pressed the camera's motor button, a large group of high-ranking officers and a couple of men in civilian clothes came in. In the center of the group was a rather short man with a broad face and slightly wavy black hairs. He was wearing the most outlandish dark-gray costume, which I took at first for a space suit like the ones in science fiction movies. In his left hand he carried a white helmet. My initial thought was that this was a cosmonaut. It was 1960, one year before Yury Gagarin's first space flight and only two and a half years after the launch of the first Sputnik. Rocketships and outer space were the topic of the day on radio and television, in newspapers and in the movies, and teenagers like me naturally dreamed about space adventures. It was no surprise that I took the man for a cosmonaut.

The officers led the man in the "space suit" to a chair at the center of the table, right in front of the microphones, and ordered him to sit down. The man seemed not to understand at first, then lowered himself into the chair without saying a word. The highest-ranking officer there, a major-general, began asking him questions. I was too busy working my camera and signaling to the soldiers who were manning the floodlights to pay much attention to what was actually being said. But then I realized that the "cosmonaut" spoke no Russian. One of the men in civilian dress was translating the questions into English. It seemed the strange man was an American.

The general, in a kind of pathetic voice, was saying, "... and we shot you down with the first rocket!"

When I heard this, I was shocked. "My God," I thought, "they've shot down an American spacecraft!"

I soon realized that the general was talking about an airplane and that the man was not an astronaut but an American pilot. The general asked him his name and the man answered. By the next day it would be a name known throughout the world. The man was Francis Gary Powers, and the downed plane was his Lockheed U-2 surveillance aircraft. The meeting I was capturing on film was the first interrogation of Powers after his parachute landing near Sverdlovsk. Following his arrest, he had been brought to the Staff building for a brief interrogation before his captors transferred him to Moscow. Evidently, the local generals wanted to learn whatever they could from him. Having such a rare bird in their hands was the opportunity of a lifetime and one they did not want to miss.

Gary Powers after his arrest

I understood from the questions that Soviet air defense systems had been tracking the U-2 for some time, almost from the southern border, but that the decision to shoot it down had been made by Khrushchev himself. Later, I learned that the reason the U-2 had not been shot down sooner was that Soviet anti-aircraft missiles lacked sufficient range to make an effective strike. In addition, the Air Defense Command feared the plane might turn back upon detecting the fire. This would have been even worse, for the unharmed plane would then have not only determined the positions of Soviet missile sites but also proven the inefficiency of the Soviet air defense system. American spy planes had been flying over the Soviet Union since President Eisenhower first ordered them in 1956, but the Russians, unable either to prove it or to shoot the planes down, had kept a low profile. But in 1959 the Soviets successfully tested the new MIG-21F interceptor planes equipped with the R3F air-to-air missiles. That new plane could reach up to 60,000 ft.—close to where the American spies used to fly. Also, in the early 1960, ground-to-air missiles were modernized to reach higher altitudes.

During the May Day parade in Moscow, Khrushchev had been standing atop the Lenin Mausoleum in Red Square when he received the urgent report that an American spy plane was again flying over Soviet territory. The plane was flying northward from Peshawar in Pakistan and had already passed over the top-secret Soviet cosmodrome at Tyura-Tam, officially known as Baikonur. Khrushchev immediately called Pyotr Grushin, the principal designer of the Soviet anti-aircraft missiles, and demanded to know why his rockets were useless for air defense. Grushin told him that in fact the rockets' top ceiling had just recently been increased and that they might have a chance.

Hearing this, Khrushchev then and there, from the podium above Lenin's glass coffin, gave the order to launch the rockets. Grushin immediately left the parade and headed for the Moscow Air Defense Headquarters to personally supervise the launch. About an hour later, Soviet Marshal S. S. Biryuzov ran up onto Lenin's tomb and whispered in Khrushchev's ear that the enemy plane had been downed and that the pilot had survived and was even then about to be interrogated in Sverdlovsk.

Of course, I knew none of this as I was watching the interrogation of a man in a dark-gray overalls through the viewfinder of my camera. Powers' replies were so terse that even I with my nonexistent knowledge of English could understand him, for the only two words he spoke were "yes" and "no." He sat quietly, looking straight ahead. He seemed almost to be oblivious to what was going on around him. Only once, at the very beginning, did he glance straight into my camera. His face was pale, and I could see that this show of calm and dignity do not come easily to him. I was especially impressed by his strength and self-control given all he had been through—having his plane shot down and bailing by parachute, landing in a plowed field, then being arrested by the local militia and transferred to the military authorities.

The interrogation was relatively brief, twenty or twenty-five minutes at the most. I filmed it in its entirety, except for a few minutes when I had to change the cassette on my camera. Finally a young officer came into the room and told the general, "We're ready." The general stood up and ordered the guards to escort Gary Powers out of the building. He was to be taken to the airport and flown to Moscow.

When we were done filming, Boyko collected the cassettes with the exposed film and gave them to the captain who was supervising our work. He explained that the military would develop the film itself, and that my part was therefore over. The very next day, however, Boyko brought two reels of the developed film to the studio and told me to start compiling it into a newsreel. Meanwhile, he and another officer went off to the country site where solders were still gathering the debris from the downed U-2, which was scattered over a large area. The soldiers were combing through virtually every square inch of field, forest, and marsh.

Two days later, Boyko took me with him to shoot a towering pile of broken metal, wires, tubes, and glass that sat in a plowed field near a small village. The area was surrounded by tight security. The soldiers searching the area were still bringing bits and pieces of the crashed plane and tossing them onto the pile. We filmed this painstaking work, then drove back to the House of Officers. A few days after this Boyko left for Moscow where he was allowed to film Powers during the interrogations and trial. He was also permitted to get some close-up shots of the most interesting equipment from the plane—photographic cameras, avionics, maps, and even some spy paraphernalia, including the suicide pin which, as Boyko explained to me, Powers was supposed to use to kill himself to avoid being captured alive. Apparently, he was unable or, in the end, unwilling to do so.

It took me two weeks to put together the news report from over a thousand meters of film we had. It contained footage I had taken during the May Day interrogation along with that which Boyko and other cameramen had shot both in Sverdlovsk and Moscow. The finished newsreel was quite good and was shown not only in all the military installations, but in some civilian theaters as well. A grateful Boyko decided to give me a present, bringing me a fragment of the U-2 plane as a souvenir. That twisted piece of dark gray metal about ten inches long stood on my desk for many years.

While editing the film, I cut negative frames from every interesting scene for my personal archive and printed photographs from them to impress my friends at school. Similar still photographs were made by the Soviet propaganda agencies from the same scenes and published in many newspapers and magazines, though, naturally, without crediting me as the photographer. When I left the Soviet Union, it was too risky to take these negatives with me. I left them behind and they vanished forever. Recently, though, between the pages of a book I brought with me to America, I found a single negative of the pile of wreckage from the U-2 sitting in the plowed field.

Pile of wreckage of the U-2 plane

Seven years had passed since those memorable days. I was about to graduate from the radio-electronic department of the Ural Polytechnic Institute. In the Soviet Union, military service was compulsory. However, students of the technical universities were exempt from the draft provided they underwent military training while at school. So it was that once a week for five years I studied air defense equipment. My military specialty was ground-to-air missile guidance. The course was very intense. We had to memorize every last detail of a guidance radar station and of anti-aircraft rockets. I spent hours studying the thick manuals for the SNR-75M station (the Russian abbreviation for "*Rocket Guidance Station, 75th model, Modernized*"). The manuals were classified as secret, though the wags among us liked to say that Allen Dulles' mother used to read him these manuals as spooky bedtime stories.

At the end of the training course we were to receive the rank of lieutenant (reserve), but first we had to spend a month in a camp with the actual guidance station and missiles. In June 1967, by an odd twist of fate, I was assigned to undergo my final field training in the same air defense division that on May 1, 1960, had shot down Francis Gary Powers.

There were still a few officers in the division who had served at the time of the 1960 incident. I talked to them, and they told me about some unknown circumstances surrounding the downing of the U-2. Like all Soviet military installations, the rocket division had a "Lenin Room"—a sort of small library of Communist propaganda literature together with a small museum containing memorabilia pertaining to the history of that military installation. On one wall of the Lenin Room hung a large bulletin board. Not surprisingly, its subject was the successful and remarkable May 1 rocket launch. I even found a couple of still photographs there taken from my film negatives.

At the top left corner of the display I saw a picture of a young officer with a dark mustache. A sign below the picture identified it as a portrait of a Captain G. (he had a Georgian name which I have unfortunately forgotten). One of the officers explained to me that this Captain G. was a casualty of the May Day incident. Initially, the decision was made to try to force the U-2 to land,

and G. was sent up in a MIG-21F interceptor with orders to make Powers come down. But the U-2 was flying at an altitude of some 80,000 feet and the interceptor could not get up that high. So G. was instructed to return to base and clear the airspace immediately, because the anti-aircraft V-750 missiles (known in the West as SA-2) were ready for launching. Each military plane and each rocket had special avionics equipment designed to identify a plane as "friend or foe" and avoid mistakenly shooting at a wrong target—in other words, to prevent friendly fire. But this time, apparently, there was a code mismatch between the interceptor and the guidance station, and the very first missile that was launched chased the poor MIG down and hit it dead on. Captain G. was killed instantly.

After this tragic accident, two more V-750 missiles were launched, yet they were also unable to reach the U-2's altitude. Ground-to-air missiles were programmed to self-destruct if they missed their target, and so, when the rockets failed to reach the plane, they exploded in mid air. But the explosion turned out to be a stroke of luck for the Soviets. Structurally, the U-2 spy plane was very light and quite weak. Though it could cover long distances at very high altitudes, it could also be easily damaged by little more than a strong wind. Flying such a craft was a very dangerous business. In this instance, when the Soviet rockets self-destructed, the shock wave buffeted the U-2 so severely that it broke apart. Any other plane would certainly have survived, but not the U-2. The rocket guidance officers at the radar station were very surprised indeed to see from their monitors that the enemy plane had broken up a fair distance from the explosion. Gary Powers had managed to get off the cockpit (the U-2 plane did not have an ejection seat), bailed out, and landed safely in the plowed field. The rest became history.

Gary Powers was sentenced by the Soviet courts to ten years in prison, but in 1962 was exchanged for the Soviet spy Colonel Rudolf Abel, captured in New York in 1957. Upon his return to America, Powers was treated very badly for letting himself be taken alive and for embarrassing the Eisenhower administration by admitting that his mission had been far from innocent. Yet even though he had been provided with a suicide pin, he had never been under orders to take his own life to avoid capture. And as far as I know, while in the Soviet Union he behaved with dignity and pride and served his country well. Still, many politicians in the United States could not bring themselves to forgive him for having survived. In October 1963 Powers married Sue Downey and the couple moved from Virginia to California. There Powers worked as a test pilot for Lockheed, though in fact his salary was secretly paid by the CIA.

Another ten years had passed, and my wife Irena and I and our infant son Roman left Russia for good. At the beginning, we lived for several months in the Austrian capital of Vienna, where I earned my first U.S. dollars as a consultant to the CIA. When I told a CIA operative there the story of how I had filmed Gary Powers seventeen years earlier, he remarked that life was full of coincidences. In the middle of May 1960, it turned out, this operative was present during Eisenhower's meeting with Khrushchev in Paris right after the downing of the U-2. He recalled a furious Khrushchev screaming hysterically, "Unless Eisenhower apologizes, I'll return home!" And he did.

The CIA operative then asked me if I would like to meet Gary Powers when I got to the United States. He told me that the former U-2 pilot now lived with his wife, son and a stepdaughter near Los Angeles and was working as a Cessna pilot for radio station KGIL, reporting on traffic conditions. Naturally, I was delighted. I doubted that Powers would remember my face, but he was the first American I had ever met, and I had been a witness to probably the most tragic moments of his life. The operative promised to see that a meeting was arranged once I arrived in the United States.

This conversation took place on July 30, 1977. Three days later, when I showed up at the U.S. embassy for my next scheduled meeting, he told me, "I have sad news for you. Yesterday, Gary Powers was flying near Santa Barbara and his plane crashed. This time he did not survive."

It may be appropriate here to give a brief account of Colonel Rudolf Abel, the Soviet spy who was exchanged for Powers in February of 1962. Though the episode has nothing to do with me, for I never met Abel, I happened to learn some curious details about his story. Some additional information I received from Marshall L., who in 1957 lived in Brooklyn across the Abel's studio No. 505 at 252 Fulton Street. Indeed, it was Marshall's apartment that the FBI used to conduct its surveillance of Abel's activities for the several weeks leading up to his arrest.

The story begins with the Spanish Civil War of 1936-39. Soviet subversive operations on Spanish soil were controlled by the commissar (an earlier rank equivalent to a general) Alexander Orlov. Orlov was a top intelligence man and knew personally a great many Soviet secret agents based in Germany, Switzerland, Great Britain, the United States, and all around the world. During Stalin's Great Purge, virtually all the top Soviet military and civil officials were executed. When Orlov received orders from Moscow to appear for a "conference" on board the Soviet merchant ship "Svir," anchored at Le Havre, he understood that his time had come. In spite of his tremendous achievement— Orlov managed to steal the entire Spanish reserve of gold and ship it to the Soviet Union, he had no illusions about Stalin's "gratitude."

Knowing full well what Stalin had in mind, Orlov decided to flee. With his wife and their terminally ill daughter he escaped to America. There he contacted the State Department, hoping that his unique knowledge of Soviet subversive operations might be of interest to the U.S. Government. Strange though it may seem, it was not. The CIA did not yet exist at this period—indeed, America had practically no foreign intelligence to speak of, and cared little about goings-on in Europe or, for that matter, anywhere else. No one in the United States was even in a position to receive Orlov's information, much less make use of it. So Orlov decided to lay low until the time was right. He spent the late forties and early fifties living in Cleveland under assumed names and working on his book *The Secret Story of Stalin's Crimes*, which he published in February of 1953, few weeks before Stalin died.

Orlov's book sent a shock wave straight to the core of the Soviet intelligence community. At the time, the KGB operations in the United States were essentially paralyzed following the arrest of the "atomic spies." The KGB wanted to reactivate a number of agents who had been pulled out of the field for "conservation" after Orlov's escape, but it was too much of a risk to call them

The First American

back to duty until the KGB knew for certain whether Orlov had "sung," unmasking those agents he knew personally. It was imperative to find out whether Orlov had in fact given up the KGB's entire U.S. network, or was still a loyal Communist and not a traitor to KGB interests.

Colonel William Fischer (Rudolf Abel) in the late sixties

To solve the mystery, the KGB devised a subtle plan. They decided to put into action in America a man Orlov knew very well. This man was William August Fischer, a German by birth but a longtime Soviet secret agent in the Western countries. Once he arrived in America, however, Fischer assumed an alias of Emil R. Goldfus. He spoke perfect Russian, German, Polish, Yiddish, and English, the last in Scottish, Irish, Oxford, and Brooklyn accents. The KGB's plan was to use "Goldfus" as bait. His mission was to start behaving suspiciously so as to trigger the interest of the FBI, eventually get himself arrested, and then "confess" that his name was Rudolf Abel. The real Abel had also been a KGB agent and a close friend of both Orlov and Fischer, but had died in 1957 in Moscow and lay buried in the German cemetery. If arrested as Abel, his true identity would be revealed very quickly—but only if Orlov spoke, for Orlov was the only man in America who knew that Abel was actually Fischer.

The plan worked perfectly. Colonel "Rudolf Abel" was arrested by the FBI in Brooklyn, where he worked as a freelance photographer. His apartment was full of spy-related equipment and code books. He was sentenced to thirty years in federal prison in Atlanta as Rudolf Abel. Curiously, the only evidence against him was the abundance of spy paraphernalia in his home, though, as far as I know, he stole no American secrets nor ever even attempted to do so. His true identity was not discovered. This signaled to the KGB that Orlov, who by this time was living in Ann Arbor, Michigan, was still keeping his best secrets to himself.

Colonel "Abel" spent over four years in an American prison, then, on the order of Kennedy administration, was brought to Germany and exchanged for Francis Gary Powers.

3

The Mind Readers

Now and then, in countries around the world, clairvoyant men and women of unusual perception appear. For lack of a better name, they are often called psychics, even though such people may be endowed with quite different talents. In Russia, mystics were always respected and held in serious regard. Some even left noticeable traces in history, such as Grigori Rasputin. Others remained mere entertainers. Virtually everyone in my former homeland knew the names of Rosa Kuleshova, Wolf Messing, Ninel Kulagina, Nikolai Kuny, and Ben Benditkis. Of them all, Messing may have been the most celebrated.

Wolf Messing was born in Germany at the turn of the century. As he wrote in his autobiography published about thirty years ago, he discovered his ability to read other people's minds at an early age. What was more, he claimed that he could influence another's mind from a distance and even subdue that person to his will. In one story he describes how, as a teenager, he was riding the train without a ticket. Suddenly, he saw the conductor coming down the aisle checking tickets. Wolf tore off a corner of a newspaper he found under the seat and gave it to the conductor, all the while willing him to believe that the scrap was actually a ticket. And indeed, when Wolf handed him the blank bit of paper, the conductor punched the "ticket," gave it back to the boy, and continued on through the car. Wolf Messing claimed that this was when he recognized his mental power—the power not just to read other people's minds, but to influence them as well.

In another story Messing told of being arrested by the Nazis for not wearing the yellow Star of David that German law in the thirties required Jews to wear. The police tossed him into a prison cell and locked him up. In the middle of the night, Messing concentrated all his mental power on making every guard in the prison gather at his cell. And it worked—they all obeyed. Again, without saying a word, he mentally ordered them to unlock his cell and enter it. They complied like robots, and he locked them all up and escaped. Heading East, he crossed the border into the Soviet Union. There he was promptly arrested by the Russian border guards. Soon, however, the stories about Messing reached Stalin, who ordered his release and granted him asylum.

For a time, Stalin himself even patronized the refugee with the unusual abilities. One day he called Wolf on the telephone and invited to come to him in Kremlin, "The pass will be waiting for you at the Borovitskaya gate". To that Messing replied, "If you want to see me I will be at your office on time and I don't need any pass". Stalin had ordered to double security and not let anyone to his office. In spite of that, Messing came straight through and nobody stopped him. During World War Two, Stalin often asked Messing what was on Hitler's mind and Messing was able to tell. There were rumors that Hitler had his own parapsychologist who tried to counteract Messing's mental power. This, at least, is Messing's version of events, though I am not prepared to bet that all these events ever really took place, at least not in the way he described them. In any event, Wolf Messing became a great celebrity in Russia. He traveled all around the country with shows he called "Psychological Experiments." Never having seen one of his shows before, I was terribly excited when it was announced that he was coming to our city. This was the spring of 1968.

Getting a ticket for "Psychological Experiments" was next to impossible. Everything was sold out months in advance. Even so, my mother somehow managed to come up with two tickets. She decided to give one to me and the other to my grandmother. At seventy-four, my granny was a very skeptical old woman. Whatever you showed her was met with a sarcastic smile: "Do you think I've never seen anything like that before? I saw all that crap ages ago, during the time of the tsar." She was an odd woman, my grandmother. So my mom decided to send her to the show, in hopes that Wolf Messing might impress her with something she had never seen before. Even during the time of the tsar.

We arrived at the Philharmonic Concert Hall early, but already there was a large crowd at the entrance. The show was sold out but people were still hoping to get spare tickets. We had excellent seats, in the center of the fifth row. The hall was packed. The show began with a dull-looking woman, Messing's manager, coming out onstage. She was about fifty-five and her entire appearance resembled a gray kangaroo. She stood there crossing her short hand on her belly, looking over our heads with gray emotionless eyes. In a monotone voice, the woman gave a formal introduction and a "scientific" explanation of what we were about to witness. As Messing told me later, when we became friends, the authorities required such an introduction so as to create a "materialistic foundation" for the subsequent experiments.

The woman-kangaroo told us that what would appear to be an act of mind reading was just an illusion resulting from the extreme perception of Messing's senses and his exceptional ability to read body language. Following the introduction, Wolf Messing walked briskly out from the wings. He was in his late sixties or early seventies, shorter than the average, with curly silver hair and a wrinkled face. Speaking Russian with a thick German accent, he told us a little bit about himself and the experiments he was about to perform. He spoke with great passion, in a quavering voice and using a number of strange words and phrases. It was quite clear to me that his speech was not intended simply to pass information to the audience, but was also a kind of psychological preparation—a warm-up to put the audience and himself into the appropriate emotional mood.

He explained that during the experiments he would act under the mental guidance of an "inductor." That is, someone from the audience—the inductor—would give him mental commands, and he, Messing, would do whatever he was commanded to do. He would perceive the mental orders of the inductor by holding his or her wrist. These orders might be of any nature, though they should not involve writing, drawing, or saying anything specific. Apart from this, the inductor could give whatever commands he or she wanted. To supervise the experiments, the audience was to select a "jury" to sit on the stage and verify the accuracy of his performance. Five volunteers were picked to form the jury, and Messing's manager then asked the audience to come up with tasks for the experiments. We were to write down our assignments on sheets of paper, then seal them in white envelopes handed out by the manager and pass them up to the jury.

Naturally, I wanted to come up with a particularly interesting problem. On my sheet of paper I wrote: "Go to the fifth row, seat 17, take the purse from the hands of the old woman, open it, find a notebook and two pencils—red and blue. Take all these to the stage. Put the blue pencil in your own left pocket and with the red pencil write on page 11 of the notebook the Latin phrase *errare humanum est* [to err is human], then tear the page out and give it to the redheaded woman in the jury." I sealed my note in the envelope, wrote my first name on it, and passed it up to the stage.

The foreman of the jury selected an envelope at random from a batch of about fifteen or twenty and announced the name of the person with the first task. The man whose name had been called came up to the stage. Messing grabbed him by the wrist and shouted loudly, "Think! Think harder!!" He began rushing with him up and down the aisle. Messing's face turned red, his hands shook, his voice became very strained. His entire body radiated an enormous tension, if not physical pain. With a trembling hand he took a comb from the pocket of his inductor, walked to the first row, combed the hair of a man in the far right seat, then rushed to another man three seats to the left, took a wallet from his pocket, removed several bank notes from it, turned them upside down, put the money back into the wallet, and handed the wallet to the inductor. That was it. The jury opened the envelope and confirmed that everything had been done exactly as written. Messing, delighted, smiled broadly and gave a bow. Apparently, he needed a few moments to rest, and he paused to recount some stories about his life, many of which I already knew from his book of reminiscences. At times it was difficult to understand what he was saying because of his accent and the use of so many Latin phrases.

After a few more such experiments, my name was called. I went up on stage and Messing immediately grabbed my right wrist with his trembling hand: "Think harder!" He yanked my hand back and forth, apparently in order to sense my involuntary reaction—whether I resisted or gave way in a certain direction. Finally, he got the response he was looking for and pulled me down from the stage into the audience, and then toward my grandmother. Jerking my hand about in every direction, he grabbed hold of her purse and took out the notebook and the pencils. Then, as I mentally instructed him, we went back up onto the stage, where he opened the notebook to page eleven and began drawing lines on the page with the red pencil. But he could not write anything and his tension grew dramatically. His mouth opened, sweat poured down his cheeks, his body shook. I could see that he was not well and could not write anything. This went on for a couple of minutes. Gently, I tried to help him by guiding his hand so that he could at least write a word, but he could not do it.

Suddenly, I remembered that before the show started, Messing had warned the audience that our assignments for him should not include writing or drawing. I felt guilty. It was clearly my mistake, so I firmly removed my hand from his grip, walked to the edge of the stage and said, "Ladies and gentlemen, I want to stop this experiment and apologize. I did something I shouldn't have done. Contrary to the warning, I commanded Wolf Messing to write down a Latin phrase. He couldn't write it. However, the phrase, *errare humanum est*, has already been heard today. Wolf Messing said it himself just before the experiment started."

Everyone was happy with my explanation and I thought the case was closed and I could go back to my seat. But Messing's reaction was quite unexpected and very emotional. Grabbing both my hands to keep me on the stage, he addressed the audience: "I want to tell you something about this young man. I have never seen him before, but I can read him and that sweet lady who I believe is his mother like open books. Let me describe them to you."

Apparently, he took my grandmother for my mother—not too good for a mind reader. Then, in front of the hundreds of people, Messing began to describe what a great woman my mother (that is, my granny) was, that she had lived a very difficult and bitter life, but that her will power and bravery had helped her and her family to survive. (This was true—not bad for a mind reader.) Needless to say, the old lady was very pleased and quite impressed.

Turning then to me, Messing said I possessed a strong mental power equal only to his own. (This was not true.) He went on about me for another few minutes. I no longer remember exactly what he said, but his entire speech consisted of compliments and fortune telling. For example, he said that I would travel a great deal and see many countries. Such a prediction in the Soviet Union in 1968 sounded like a fairy tale, to say the least. It was a time when ordinary people like me were not allowed to travel freely even within the country, much less abroad. Oddly enough, just nine years later I managed to get out of the Soviet Union, and as of today, I have already traveled to over thirty countries. At the time, however, I was convinced that Messing was saying those things because he was grateful to me for helping him out of an embarrassing and even painful experiment.

I suspect that many participants in his shows tried to catch him and expose his errors. I did not want to do this for the simple reason that the stage in not the right place for any serious experiment, psychological or other. I thanked Messing for his kind words about me and went back to my seat. When I was leaving, he whispered in my ear, "Please, come see me after the show. I want to talk to you."

When the performance was over, I walked my grandmother outside and asked her to wait for me. She did not mind at all, as a large crowd had already flocked around her. Everyone wanted to know if Messing had been accurate in his description of us. It was probably the first time in her life that she had enjoyed such pleasant publicity.

When I got backstage, Messing was surrounded by a dozen or so women and children. Evidently, these mothers had all brought their troublesome offspring to him in the hope that this man with such "un-human" power could miraculously cure them of the stubborn diseases on which traditional medicine had given up. This was a rather common routine after all his shows.

I had to wait about fifteen minutes before he managed to free himself. Coming over to me, he said, "You don't realize it, but I see in you my perfect inductor. I can communicate with you much more easily than I ever could with anyone else."

In reality, Messing was quite convinced that he indeed possessed telepathic abilities and ESP. I, however, tend toward a much more prosaic explanation of our "super-communication." During the show, as I was gently moving his hand in an attempt to help him write a few Latin words, he took my help for some unconscious mental communication between us. No doubt, all his life the people he brought on stage tried to prevent him from succeeding, and never wanted to help. Alas, from me he received more than he ever expected.

Then he asked me, "Would you like to take part in some mental experiments with me?"

"Of course", I said. "I'd love to. But how?"

"I will send you my photograph. At a specific time, you must look at the picture, recalling my face, and send me a mental message. Any message you think of. I may be thousands of miles away, but at that time I will be checking my sensations. Later, we will talk on the telephone and compare our results."

I did not think it was a scientifically well-designed experiment, but not wanting to disappoint my newly found "inductor," I agreed. Messing gave me his address and home phone number and then asked me for mine. When I wrote my last name into his notebook, he suddenly exclaimed, "Oh, my God! I knew it! I felt it! Now everything fits together perfectly. It's wonderful! It's fate! Oh, my Go-o-od!"

"What's the matter?" I asked, quite puzzled at his emotional outburst.

"Your name! I knew it!" He was shouting in his excitement. "You see, forty years ago I lived in Vienna and studied with the great Sigmund Freud. And your name is Fraden!" (In Russian, the pronunciation of these two names are very close: "Freyd" and "Freydin.") "No, it's not a coincidence," he went on. "It's fate! You, my best inductor, have the same name as my beloved teacher! I felt it the very moment I saw you!"

Apparently, the similarity in our names thoroughly convinced Messing that there was a direct spiritual link between Freud, himself, and me. I had no wish to discourage him (and even if I had tried, he would not have believed me anyway) and promised that the Wednesday after receiving his photograph, at 3:00 P.M. local time, I would send him my first message, and that same evening would wait for his telephone call.

His grim-faced manager came over to us, put her hand on Messing's shoulder and said, "Wolf Grigorievich is tired, he needs rest after the show. We must go. Good-by." Indeed, he looked completely exhausted and worn-out. Wishing the two of them good night, I rushed out to find my grandmother. It had been an interesting evening for both of us.

Wolf Messing, the celebrated mind reader

A couple of weeks later, I received the photograph and a letter from the northern town of Vorkuta, where Messing was giving his show. The letter gave me instructions on how to concentrate while mentally communicating with Messing. He wrote, *"Also, I am asking you to be absolutely positive of your abilities. I sense them in you and believe in them. My best regards to your mother."* The photograph was signed, *"Mentally I am with you. Wolf Messing."* To be true to my word to him, I did as I had promised. On the day specified, I sat on a bench in the park, staring at his picture. I tried to concentrate as hard as I could, though obviously, no matter how intensely I tried, there was no way I could communicate with him except by the telephone. Messing was not discouraged by our failure and invited me to come to see him in Moscow.

During those years, I knew a beautiful girl who lived in Moscow. I wanted to marry her and would travel to Moscow any chance I could. Eventually, we decided not to marry. As one joker put it, we broke up because of an illness—we became sick of each other. But before this, I visited Moscow at least five or seven times a year, and whenever I was there, I saw Messing. If he was not on a tour, we would meet either at his home on Novopeschanaya Street in the northern part of Moscow, or at a hotel.

One evening we met at the Rossia Hotel, at that time the largest hotel in the world, located just a block from the Kremlin. As I entered the lobby, I saw Messing standing in the center. A short plump man was whirling around him. After I greeted them and we had chatted for a bit, the plump man said in coarse voice, "Wolf Grigorievich, I am going to get some beer. I am awfully thirsty."

Messing smiled, "Go, go, Yuly. I will be waiting for you here with my young friend."

"Who was that?" I asked when the thirsty man had disappeared.

"He is a writer. He specializes in detective novels. His name is Yulian Semyonov. A very gifted man, and pleasant, too. The only problem is that he can't write, or for that matter, do anything useful when he is sober. But what can one do? After all, he is Russian."

The name was quite familiar to me. Semyonov's books were bestsellers, and he was one of the most celebrated writers in Russia at the time. Messing went on to explain to me that Semyonov was working on a new book about a fictitious Russian super-spy during World War II, a sort of Soviet James Bond. And he wanted to make his spy a mind reader, something like Messing. This was why he had come to see Messing—to get a better understanding what was it was to be able to read people like open books. The novel was eventually published about a year later, and a very successful television serial, "The Seventeen Moments of Spring," was produced based on it. The story concerned a Soviet intelligence officer named Stirlitz. After reading the book and watching the television serial, however, I never did find any similarity or even resemblance to Messing.

My contacts with Messing continued for several years, until his death on an operating table in a Moscow hospital. Sometimes we met in person; other times we communicated by telephone, or occasionally by mail. I still have his letter and the photograph, and a telegram from him with warm New Year's wishes. I always sent him greetings every year on his birthday, September 11. (Coincidentally, it is September 11 as I write these lines.)

Messing was a very kind man. He was forever amazing me with his brilliant wit and insight. Once, I met him in Moscow the night before flying back to Sverdlovsk. The next morning, after arriving home, I went to the medical school to hear a lecture on psychology given for the medical students by Dr. Zorin. Following the lecture, one of the students asked Zorin a question about Messing's ability to read minds. Zorin began his response by refering to "the recently deceased medium Wolf Messing." I was shocked. I had just seen Messing the previous evening. He had been fine. What was going on? Had Zorin simply heard a false rumor that he had died (there were all kinds of rumors about Messing), or could it possibly be that Messing had indeed died during the night? I rushed to the post office (in those days to make a long distance call in Russia one had to go to a post office), and dialed his Moscow apartment.

The ever-present woman manager (who managed his life as well as his shows) answered. I told her who it was, and Messing picked up the phone. Before I could say a word, he asked, "What's the matter? Why are you calling? If you think I am dead, you are quite wrong! I am perfectly okay, but thanks for worrying anyway. Good-by."

Could he really have read my mind, from a thousand miles away?

It is a pity that he was wrong about me. I do not have his talent for reading other people's minds, and people sometimes deceive me. I was never able to develop such sharp senses in myself, could never "read people like open books." But I did learn how to recognize if a person has such abilities. From Messing I learned something that helped me to understand better many "facts" of parapsychology, ESP, and the like.

Psychics have an extremely fine-tuned perception of the physiological reactions and the body language of their inductors. This is why we call them extrasenses or mediums. An expert with a polygraph (a "lie detector") can record and analyze involuntary and generally unnoticeable eye movements, perspiration, changes in heart rate, skin resistance or muscle tension. In a similar way, the extrasense can perceive these same changes— only without any apparatus, just through his or her own natural sensations. In most cases, the medium does not even realize how he does it. It just happens. He unconsciously feels the other person's response. To make this easier, he may hold the inductor by the wrist, just as Messing did during his "Psychological Experiments."

One day, I received a call from a local newspaper. They had heard about my contacts with Messing and told me that another medium was coming to town. He would be doing a similar, private show just for the media and closed to the public. The man from the newspaper asked me whether I might come observe the show, and later shed some light on it.

In the editor's large office, the newspaper journalists gathered to meet the visiting psychic, Benditkis. The room was packed. Some were sitting in chairs they brought in for that occasion, some just stood around leaning against the walls. About fifty years old, the psychic was a tall, good-looking, and well mannered man. He spoke to the journalists quietly and intelligently. His experiments were similar to those of Messing, yet his style was calm and serene, without the nervous tension that typified Messing's performances. The experiments were quite impressive and thoroughly successful, leaving everyone virtually dumfounded.

I was sitting in a corner of the room. The journalist who had invited me whispered in my ear, "Why won't you participate? Can you do something interesting?" I decided to give it a try. Toward the end of the show, Benditkis said that he could read an inductor's mind even if he could not see or touch him. He asked us to hide something small, such as a pen, a button or a matchbox, while he waited outside the room. Then, someone from the audience was to act as an inductor, standing behind Benditkis' back and "thinking" to him where to look for the hidden object.

One of the journalists gave him a small gold-plated pin. When Benditkis left the room, I proposed to the others that I hide the pin on my person. I hid it on my head, in the hair above my right ear. I then went back to my chair and Benditkis was called back into the room. One of the younger journalists volunteered to give Benditkis mental commands. He stood behind the medium, and Benditkis told him, "Think hard directly toward the back of my head. Look straight at my head and think about where the pin is hidden."

The naïve young fellow tried really hard. Benditkis faced the audience with the young man standing behind him. The psychic slowly scanned everyone's face until his gaze rested on me. He approached the corner where I was sitting (the "thinker" at his back following him closely) and invited me to step forward. I came to the front of the room and stood facing him with my back toward the audience. Benditkis began searching, stretching his fingers toward me but not touching me. His hands moved up and down from my shoes to the top of my head, to the left and to the right. I realized that this had nothing to do with the "thinker" behind his back. Benditkis was reading *my* body language, not young journalist's "telepathic waves."

"Aha!" I said to myself. "You are playing a game with us. Well then, I'll play a game with you." And I started giving him wrong messages by simulating involuntary reactions. Since the pin was hidden above my right ear, whenever he moved toward my left side, I would either blink, or hold my breath for a beat, or shrug almost imperceptibly, or something of the sort. He searched and searched and searched over the left side of my body—and, of course, found nothing. Whenever he was turning toward my right side, I was mentally relaxed.

The poor psychic worked hard for several minutes, finally beginning to panic. His successful show was ending in total failure. Finally, he threw up his hands and said that he was awfully tired and could not complete the experiment. The show came to an end. I decided not to tell him about my little trick. After all, he had lied about sensing by the back of his head, so I did not feel too bad about ruining his show.

In 1967 I received a master's degree in electrical engineering and decided to work toward my Ph.D. in medical instrumentation. In Russia and in some European countries this is called the Candidate of Sciences degree and does not necessarily require attending postgraduate school. It did, however, involve undertaking a serious research project with thesis to be defended before an exacting panel of scientists. Usually, such a project took several years of hard work under the guidance of an advisor who must himself possess the highest scientific degree in Russia: Doctor of

Sciences (Sc.D.). This degree really has no equivalent in America and as a rule was granted to individuals who had not only earned the Candidate degree but had also made significant contributions in their particular scientific fields.

On January 2, 1968, I began my professional carrier at a medical electronics laboratory affiliated with a research institute specializing in occupational diseases. After about a year there, I began looking about for an interesting problem for my Ph.D. research project. I could, of course, simply select some engineering area and zero in on one of many unsolved questions in medical instrumentation. But I was young and wanted something more exciting, unconventional, and enigmatic. At the time I was still meeting frequently with Messing, learning from him how psychics perform their tricks—but this was all on the side. Obviously, it had nothing to do with my work and studies. After all, I was an engineer, not a psychologist.

One day, my boss and research advisor, Professor Vladimir Rosenblat, M.D., Sc.D., told me that he was planning a physiological conference to discuss paranormal phenomena. He said he had made arrangements for the famous psychic Rosa Kuleshova to give a demonstration. For the past eight years or so Kuleshova had been the subject of a number of studies by two doctors in her native town of Nijny Tagil, about fifty miles from Sverdlovsk. Rosenblat knew Dr. Yakov Fishelev, one of the doctors who had been experimenting with her. Dr. Fishelev had grown very frustrated with his investigation. His research was going nowhere and it was clear that he could never earn his Sc.D. degree with it, as he had hoped. In the end he had given up on Kuleshova. He asked her to come to our conference and feel free to do anything she liked, since he no longer cared about her.

I was very excited. I had heard many interesting stories about the woman, but had never seen her in person. During the sixties, she was a household name not only in Russia but throughout the world. Everyone knew about her ability to read printed text by the palm of her hand, without even touching the paper.

In December of 1968, about a hundred people gathered for the conference, most of them medical doctors. I invited a few of my engineering friends as well. We sat in the auditorium of the research institute and listened to presentations by medical doctors and biologists. They spoke on parapsychology, ESP, telekinesis and other strange phenomena. Yet since the speakers themselves had no clear understanding of what they were talking about, all the speeches struck me as scientific-sounding rubbish that made little sense. Rather like a smoke screen to conceal their lack of knowledge.

The main problem with all these paranormal phenomena is the enormous difficulty and, for the most part, impossibility of reproducing them under controlled conditions. Modern scientific methods dictate that any observation must be reproducible within specific conditions. When it is impossible to control all the conditions, statistical methods are used. But it is imperative that one be able to repeat the same experiment over and over again and get identical results. If a phenomenon is not reproducible, or one is unable to repeat the experiment, it is not science. At least, not science as we think of it today.

Of course, if something is not science, it is not necessarily bad. There are a great many wonderful things which are not science—for instance, music or painting. Science is just one good thing among many. I enjoy watching and even participating in paranormal experiments in the same way that I enjoy going to a concert or a movie. Just as an actor or a writer needs his talent, so a medium needs his natural endowment. It is just a different kind of talent.

One may believe in UFOs, ESP, and so on, but knowing and believing are two different things. Faith is neither science nor art. It is not based on *facts*—be they God-made facts (science) or human-made facts (art and invention).

The big question, then, is: What is it? A science, an art, a faith? All of us sitting in that conference auditorium wanted to see some real demonstration of psychic ability. We wanted to witness *facts*. And I too was impatiently fidgeting in my seat, in anticipation of meeting Miss Rosa.

When the medical presentations had all been given, Dr. Rosenblat told the audience that today we had a special guest who was well known for her unusual ability to perceive visual information without the use of her eyes. It was a bit odd hearing Dr. Rosenblat talk about "visual information," as he himself had been blind for many years. However, we had all grown used to his incredible ability to ignore his handicap and lead a normal life both inside and outside his profession.

Dr. Rosenblat invited Rosa to say a few words to the audience before she began her demonstration of her talents. Rosa got up and immediately, in a monotone voice, began telling us about herself. She spoke quite calmly, firmly, and confidently, like a teacher attempting to impress essential bits of common knowledge into the stubborn heads of some not terribly bright school children. She referred to herself, oddly enough, in the third person: "Rosa can do this, Rosa feels this way, Rosa likes this," and so on. Twenty-eight years old, she was a plain woman with a round face, small gray eyes, and yellowish hairs, dressed in a knit jacket and brown skirt. It was clear that she was well accustomed to such meetings and to talking about herself. Her voice reminded me a robot's from a science fiction movie, with no modulation or emotion.

Rosa had with her a large album which she proudly showed to the audience. It was full of clippings of articles about her from magazines and newspapers. There were also copies of letters from a number of scientific luminaries who had witnessed her shows. She was very sure of herself and addressed everyone the same way, making no distinction between professors and laboratory technicians.

She began her experiments by reading a book with the palm of her right hand. One doctor and one lab technician carefully blindfolded her with a thick black cloth. On a table, there were several books which we had brought in from the medical library for this occasion. The doctor opened one of the books at a random page and handed it to Rosa. She placed the book on her left hand and, holding her right palm at a distance of about one inch from the page, began slowly reading. It was amazing. We were all astonished.

I wanted to take a closer look at what was going on. I went up to her and checked the blindfold. Everything was in order. Then I asked Rosa if she could put her hand on my wristwatch and tell me the positions of the hands.

She said, "Yes, I can. But don't try to catch Rosa. She can't work if people don't trust her!"

I said, "Of course I trust you. I just want to see more of your talents. Can you tell me where the hour, minute, and second hands are?"

As I said this, I turned the knob on my watch, shifting the hands to a random position. Rosa, still blindfolded, put her hand on the watch crystal, which was facing toward the floor, and said, "Ten minutes to two." She was right.

"Where is the second hand?" I asked. This she could not tell me. Perhaps because the hand was moving faster than the others. But in any event, she was doing quite well so far, and none of us could even begin to guess how she did it.

I cannot say that I smelled a rat, but I was unwilling to accept what I had seen passively, without questions. I have witnessed a few great magicians and in most cases could not guess their tricks either. But these magicians never pretended that their tricks had anything to do with the paranormal. Rosa, on the other hand, did, and this bothered me a lot. I decided to sieze a more active role in the show—that is, to suggest more experiments and ask Rosa to do what I wanted, not what she wanted. Her response, however, was swift and unconditional. She said, "Rosa refuses to work if people are skeptical. She will perform only her own experiments. If you don't like it, she will go home." So I gave up and sat passively for the remainder of the show.

The next experiment proceeded strictly according to her own instructions. It involved the detection and recognition of fingerprints, and went as follows. Rosa first asked one of the doctors to select a leather-bound book from the table and wipe off its cover with a napkin dipped in alcohol. Then, she announced that she and two witnesses would go out of the auditorium and wait for a few minutes. The witnesses would confirm that she was not peeping. While she was out of the hall, someone from the audience was to impress his or her fingerprints on the book cover, and Rosa would then tell us who it was. Once she was out of the room, one of the nurses sitting in the third row touched the book with both her thumbs and index fingers, and we lined up about ten people, including the nurse. When Rosa came back, she picked up the book and, barely touching it, began scanning the cover with her fingertips, inch by inch. Then she approached the participants and examined their finger tips with her own, one by one. This time she was not blindfolded. When she touched the nurse's hands, she said almost immediately, "These are her fingerprints." Everyone was impressed—myself included. It took me some time to figure out how she did it, but I shall explain later.

Then she said, "Rosa can also perform some medical diagnostics. Without using an X-ray machine, she can sense through a patient's stomach whether he has a stomach ulcer, cancer, or whatever else it might be."

Dr. Rosenblat said, "We can arrange that. Here in this building, we have plenty of patients of that kind. Let's take a ten-minute break and after that we will continue."

He spoke to a few doctors who agreed to bring some of their patients with stomach ulcers down to the auditorium. There was no shortage of such patients in the clinic, which treated occupational diseases. After the break, we lined up about fifteen patients, only half of whom, however, had stomach ulcers. The others were brought in for control purposes. The patients all wore hospital pajamas and slippers. Rosa asked them to unbutton the tops of their pajamas and bare their bellies. She then walked down the line, placing both her palms on the stomach of each patient for a few seconds before announcing, "This one has no ulcer, this one has one on the right side," and so on. She was right in every case. I was feeling very uneasy. There was some trick here, but so far I could not figure it out. So I asked her, "All these people have fairly large stomach ulcers. I myself have a small one. Can you tell me which side of my stomach it's on?" I unbuttoned my white coat and shirt, and Rosa slid her hand over my stomach. She touched the left and right sides and then said, "You are doing it again. You are trying to catch Rosa, but she knows that you are lying! You have no ulcer at all." Well, she was right again.

When the conference was over and Rosa was preparing to leave for the train station and return home, I spoke to my boss, asking him if it would be possible to keep Rosa there, in the city. I told him I thought it would be great if we could continue the experiments at our laboratory, and that perhaps we might discover something interesting about her. Not that I hoped to make this into a Ph.D. project—it was quite obvious that whatever we found would not be suitable for a dissertation. I was simply curious to learn more about Rosa and her experiments.

Dr. Fishelev, who was also present and heard me talking to Professor Rosenblat, said to me, "Don't be a fool. I spent five of my best years with her, and all for nothing, like a soap bubble. She's a crook. I believe that in the past she really did have some paranormal senses, but once we started working with her, she gradually lost nearly all of them. Now all she does is perform feats of circus magic. She's afraid that we may discover that she's ordinary, and so she keeps doing her hocus-pocus all the time. Just to keep us puzzled. Besides, she's a progressive case of schizophrenia. Don't do it, Yakov. [This was the Russian pronunciation of my name.] Mark my words—stay away from her."

But I did not care. If her abilities were nothing more than tricks, I had to find it out. I told Dr. Fishelev that I had no intention of spending my life trying to crack Rosa's tricks or investigating her mental or physiological gifts. I just wanted to see for myself what it was all about. He, as a doctor, had not been able to discover anything meaningful in her. Who knows—perhaps I would have more luck as an engineer and a physicist.

Dr. Rosenblat told me that he would offer Rosa a position as technician at the laboratory (in the Soviet Union, it was essential to have a job before being allowed to move to another town or city), and as soon as she relocated, we could start testing her. He then asked Rosa if she wanted to work with us, and she agreed with pleasure. Not that she was anxious to solve any scientific mysteries. I suspect she felt that the scientific community was losing interest in her, and this new opportunity was certainly better than nothing.

It was January of 1969 when Rosa joined the staff of our research institute and we began the experiments. Dr. Rosenblat, with his medical research assistants, nurses and technicians, conducted countless medical tests on her. I began by measuring electrostatic and magnetic fields. Then, using photodetectors and a photomultiplier, we tried to determine what minimum level of illumination and what photon flux Rosa needed to be able read a text with her palm. We recorded her brain waves (EEG), electrocardiogram (EKG), eye movements, temperature, and other vital signs. In addition, there were the many standard tests commonly used in studying ESP, such as cards depicting geometrical shapes—squares, triangles, circles, and so on. The testing went on for some two months, but, as Dr. Fishelev had predicted, we got nowhere.

At times Rosa was cooperative, but often she simply refused to participate. She behaved like a spoiled child. We never knew what to expect from her. Still her best trick was reading a text with her palms. No matter how many times I witnessed it, I could never understand how she did it. Like many celebrities, she was thoroughly self-obsessed. Indeed, I would even say that she was hopelessly in love with herself, and in this people have no rivals. She sincerely believed that everyone longed for nothing more than to see Rosa and the demonstrations of her talents.

On a number of occasions I caught her red-handed in an attempt to cheat or resort to circus-type magic during the serious experiments, just as Dr. Fishelev had warned us. But I understood that whatever she did was natural and candid, with no premeditated desire to deceive us. She was not nearly bright enough to fool us that easily. Her tricks or her paranormal abilities, if such they were, were unfeigned parts of her personality. She herself believed in them firmly, and could not comprehend why anyone else might doubt.

As often as we found her to be just a trickster, there were other occasions that leave me puzzled to this day. Had she lived a couple of centuries before, her natural destiny would have been to become a witch. And quite likely, she would have burned for it. Today she was simply surrounded by a bunch of curious scientists.

Her presence in the research laboratory was a cause of excitement and sometimes of disruption. One day, I saw our department secretary running down the corridor. Hot on her heels was Rosa. The poor secretary was terrified and very much confused. She hid behind my back to get away from Rosa.

"Mila," I asked, "what's this? Rosa, why are you chasing this nice woman? Would the two of you tell me what's going on?" In a trembling voice, Mila explained that she had been sitting at her typewriter working on a report, when Rosa walked in and, with no provocation or without saying a word, slipped both her hands under Mila's dress and grabbed her underwear. Mila had jumped to her feet at once and started running. And Rosa had darted off after her.

As might have been expected, Rosa's explanation was totally innocent. "What's to be afraid of?" she asked. "Rosa just wanted to see with her own *hands* what color her undies were. What's wrong with that?"

Another day she said to me, "You know, Yakov Vladimirovich, Rosa can read words not only with her hands, but with her lower behind as well. She wants to show you."

"Rosa Alexeevna," I replied, "I don't think I want to be doing *that* kind of experiment."

"There's no need to be embarrassed," she said. "Rosa will do it through her skirt. She can sit on a sheet of paper and read it like that," she said, snapping her fingers. "Only the letters must be large."

"Okay," I said. "If it's through your skirt, let's try it."

I got a bunch of large cards with different numbers on both sides and we began the experiment. We did not use a blindfold. Rosa stood with her back toward a chair, and I stood behind her, holding the stack of cards. I would place a card on the chair and tell Rosa to sit down. She would then sit on the card and immediately tell me the number on it. She got every one right. Then I decided to modify the experiment slightly. As before, I placed each card on the chair seat and told her when to sit. This time, however, just as her behind was about to touch the card, I flipped it over so that a different number was facing up.

Now she got every one wrong. It went on like that. She would sit and tell me, "It's number 5," and I would say, "No, it's not." And she would jump up, turn around, and look at the card with great surprise. We repeated it several times. She was nearly in tears.

"See here, those are trick cards. Tell Rosa what they are."

"Fine," I said. "Let's make a deal. You tell me how you read through the skirt with your behind, and I'll tell you about these cards."

She agreed and proceeded to tell me that she could not actually read through her skirt. What she did, she claimed, was smooth out the folds of her skirt folds with her hands as she sat down—and since her hands were near the card at that moment, she read the number with her palms. That was her trick, and I then told her mine. But did she really read the numbers with her hands? Was she genuinely "skin-eyed," or was she just a hoax? I still do not know for sure.

Our experiments with Rosa continued for about three months. Then she left us and went to take another job. I had no word of her for a long time, until one day the physician Sonya Gofman, who had also worked with Rosa, told me that she had seen Rosa at a bus stop. Rosa had been sobbing, telling the passersby that she had lost her most precious treasure on the bus—her album of clippings.

It is my belief that Rosa had abilities similar to those of Messing and Benditkis. She unconsciously sensed other people's involuntary physical reactions. When she was detecting fingerprints or "diagnosing" ulcers, she in fact had no perceptions whatsoever from either the fingerprints or anyone's internal organs. What she was doing was reading people's faces, heartbeats, respiration, and so on. Her inductors naturally knew whose fingerprints had been made on the book, or where an ulcer was located. By sensing people's body language, their vital signs, she arrived at the right answers. But it was obviously that herself had no idea how she did it and sincerely believed in her own supernatural abilities. It was very exciting, fascinating, and certainly amazing. But not terribly mysterious or magical.

Or was it?

4
Making an Inventor

> When I was young man I observed that nine out of ten things I did were failures. I didn't want to be a failure, so I did ten times more work.
> - *George Bernard Shaw*

*M*y first invention was probably an inkwell. I was ten years old, and in those days children used steel pens for writing (and sometimes for fighting). We had to carry portable glass or ceramic inkwells to school because no other writing utensils were allowed. I lived in the surrealistic world of a Communist party state where ball-point pens were strictly forbidden as the invention of a "decadent capitalistic society."

At home, I did my homework at a desk that had a large cast iron inkwell with a massive cap on it to keep the ink from drying out—or, perhaps, flies from drowning. Every time I went to dip my pen into the ink, I would have to open the cap and then close it again. It seemed terribly ordinary—open, close, open, close—so I decided to find a better way to manipulate the cap. "Better" meaning a way to be able to open it not by hand, but rather with my foot. Using an erector set my parents had given me for my birthday, I fashioned a series of pulleys to which I attached fishing lines connected at one end to a floor pedal made of a wooden plank, and at the other, to the cap of the inkwell. It worked great. Whenever I needed to dip my pen into the ink, I pressed the pedal and the strings moved up and down, lifting the cap off the inkwell. When I released the pedal, the cap settled slowly back down onto the inkwell, gently and precisely.

The desk where I did my homework stood in front of a large window hung with white tulle curtains. One unfortunate day, a string got caught in one of the pulleys, and when I pressed the pedal too hard, the entire assembly, including the cap and the inkwell itself, flew up as though launched from a slingshot. The beautiful tulle curtains, my mother's pride, were splattered all over with ugly violet ink. Disaster though this was, it taught me a valuable lesson: whatever one designs, especially something intended for consumer use, must incorporate adequate safety valves and be sufficiently foolproof that some idiot will not be able to push a pedal too strongly and make a mess of it. Yet, it is impossible to make everything really foolproof because fools are so ingenious!

When I was thirteen, I built my movie camera. Properly speaking, of course, it cannot really be called an invention, for the only new elements about it were my numerous makeshift solutions. Building both the camera and the film developing device was good engineering practice, requiring me to find many substitutions for parts I could neither buy nor make myself. I used a crankshaft from an old Singer sewing machine, a lens from a magnifying glass, and gears from old heirlooms. The light lock was made from a felt strip that I had cut from an old hat of my father's; a couple of dozen spools found new life in the film developing tank; and the iris aperture was taken from an old bellows-type view box camera. The project was a great exercise for the imagination—how to find new, hidden possibilities in ordinary things? It is this ability to see what other people cannot that makes an inventor.

> Everything that can be invented had been invented
> - *Charles H. Duell,*
> *Director of U.S. Patent Office, 1899.*

The Soviet dictator Joseph Stalin liked to claim that nearly all the world's greatest innovations were made by the Russians. In school we were taught that the father of the electric light bulb was not Edison, but Yablochkov. We were told that the first airplane was invented by Zhukovsky and that the first radio-receiver was built by Popov, not Marconi. For fairness' sake, however, it should be noted that Yablochkov did indeed produce electric bulbs with carbon filaments which were used for some time to light several St. Petersburg streets and even a few homes. And he did this long before 1879, when Edison finally figured out how to create a good enough vacuum within a bulb to prevent oxidation of the filament.

Professor Zhukovsky, meanwhile, came up with the fundamental theory of heavier-than-air flying machines long before the Wright brothers made their first flight. And Popov did indeed design a radio-receiver that rang a bell in response to lightning and the strong electromagnetic turbulence it produces in the atmosphere. He built his "thunderstorm detector" many years before Gugliermo Marconi and Nikola Tesla invented radio communication.

To be sure, these pioneering works were all extremely important, but still they were not crucial steps on the path to what we call an invention—the creation of a *practical* device or process. Leonardo da Vinci, arguably one of the greatest geniuses who ever lived, discovered the principle of the helicopter, but the first practical vertical flying machine was invented and built by the Russian-American Igor Sikorsky, who is rightfully recognized as the inventor of the helicopter. There is no question that the Russians have fathered some great inventions. Many of them, however, were condemned by Stalin to virtual oblivion because of having been invented by "traitors"— Russians who dared to live abroad, such as Sikorsky. In school we were taught that Pavel Nipkov developed the principle behind scanning an image in 1873, but we never learned that it was Vladimir Zvorykin, a naturalized American citizen, who in 1934 invented the iconoscope, an image-sensing device which made practical television possible.

While living in the Soviet Union, I found that the great majority of Russian inventions were of the "get-around" type. That is, they were inventions that achieved the desired effect not by attacking a problem head-on, but rather by some alternative means, which, while often neither terribly efficient nor very elegant, nevertheless allowed the goal to be reached faster and for much less money.

We rarely invented anything uniquely new. In the Soviet Union I held a patent for a method of tape recording medical information. When I came to America, I found that a similar recording device already existed and was readily available, though it was quite expensive. My tape recorder, while twenty times cheaper, was not as efficient, but it recorded with acceptable accuracy and served the purpose well. I had had to invent it because I needed such a recorder for my research projects but could not afford to buy the Japanese-made model. Another of my Soviet inventions was the so-called "sample-and-hold" electronic circuit. I was very proud of this invention, for it served as a quite useful building block for many applications in which electronic signals had to be converted from analog to digital format or vice versa. But how great was my surprise when I discovered in America that similar circuits had been available for many years. Unable to buy these chips for my projects in the Soviet Union, I had had no choice but to invent an alternative version.

Such inventions may seem like a waste of talent, but in reality it was a great school of mental efficiency and creativity. For instance, say you want to play music but the only instrument you have is a hammer. If you use it for driving nails into wood, it is not an invention—that is what the hammer is all about. But if it happens to be the only thing you have at hand, you start wondering what else you might be able to do with that hammer. Is there any way you can modify it for your purpose? Can you use it in a different manner? And, if you are lucky enough (and smart enough), you may come up with an idea for producing musical sounds with that hammer. At this point, it becomes an invention.

Many our inventions were just solutions for getting around a deficit of available parts, ways of overcoming the problems of poor equipment, low-quality components, lack of money, and, sometimes, government restrictions. Our inventions were the fruits of reevaluating the merits of ordinary objects. Forced to take a fresh look at them, we often managed to see hidden value in them. And that is how you make a good invention.

A number of Russian inventions concerned things the rest of the civilized world had no interest in doing. For years, the Russians were continually inventing and re-inventing that which already existed in better form elsewhere. In other words, the civilized world had analogous products, but the

Russians either did not know about them or could not afford them—or perhaps did not want to produce these things themselves for some reason. So, they had to invent an alternative means.

An interesting example of such creative thinking was the Soviet space program. In early 1957, the Americans were working on preparations to launch the first Earth satellite. The Soviets, however, were at that time preoccupied with building missiles to drop nuclear bombs on the heads of the "vicious American capitalists." The head of the Soviet rocket program was Sergei Pavlovich Korolev, a truly great engineer and inventor, a pilot, and a Gulag survivor. Korolev already had ballistic missiles powerful enough to deliver nuclear weapons to American airspace, but these rockets were just big flying engines with no real "brains." In the fifties, Soviet industry was still without computers, because Stalin considered cybernetics a capitalist gimmick and had forbidden any research in that direction. Hence, computer science, being a branch of cybernetics, remained hidden away in a closet. Ironically, some twenty years earlier Hitler had similarly forbidden all research in the area of electronics, since he regarded electronics as a "Jewish science." As a result, the Germans had no efficient radars before World War II (later on they did), which contributed to the failure of their attempt to invade the British Isles. Likewise, the Russians did not possess computers until the end of the fifties.

In the spring of 1957, one of Korolev's assistants brought him an English-language paper outlining the American space program. Korolev read the article through, then, after some thought, picked up the phone and called Nikita Khrushchev, the Soviet Prime Minister. He told Khrushchev that he had come up with a great propaganda scheme to convince the entire world of the Soviets' scientific and economic supremacy.

Korolev's idea was to launch the first Earth satellite before the Americans. His missile did not have the power to send a payload of true scientific significance into orbit. Besides, the lack of computers capable of performing real-time computations precluded Soviet scientists and engineers from achieving the accuracy required for a controlled flight, or for collecting any useful scientific information. In effect, nearly all essential computations had to be performed with a slide rule. (In December of 1993, Korolev's slide rule was sold at auction at Sotheby's in New York for $24,150) Still, Korolev's missile was at least powerful enough to throw a simple transmitter into orbit.

Khrushchev gave Korolev the go-ahead and a limitless supply of money, people, and equipment. The Soviet space program was born, and just few months later, on October 4 of that same year, the first Sputnik (Russian for "satellite") was launched. This first Soviet satellite was just a very simple radio transmitter operating on two frequencies, 20 and 40 megahertz, specifically selected because they were capable of being picked up by amateur radio operators anywhere on Earth. The transmitter could produce only meaningless signals, a constant "beep-beep-beep." But those "meaningless" beeps announced to the world in all languages that the Russians were in space. Korolev had pulled off one of the greatest Soviet propaganda bluffs of all time. His other great feat, however, was of a true engineering nature, and was made several years later.

Korolev knew about NASA's plans to send a man into the orbit and again wanted to beat them to it. The problem was that Soviet rockets were still a far cry from the powerful thirty-three-ton Redstone missile the Americans had prepared for the launch of the manned Freedom-7 capsule. There was no way a Soviet missile could send a man into orbit. After making some calculations, Korolev had a brilliant idea: if a single missile was not enough, why not fasten five of them together in a bundle? And this was exactly what he did. Five rockets with total of 30 engines working in concert proved capable of producing sufficient thrust to launch the first man into Earth's orbit. On April 12, 1961, the day I turned sixteen, Air Force Lieutenant Yury Gagarin became the first man in Space. His flight was made possible thanks to a "get-around" invention which worked perfectly.

Of course, nothing is easy. Bundling five rockets together created many problems, especially in terms of synchronizing their operations and stabilizing the flight. But such difficulties were minimal compared to designing a more powerful missile. Korolev's engineering solution was so effective and

Making an Inventor 77

inexpensive that such missiles are used by the Russians to this day, some thirty-five years after his invention.

The impressive ability of Russian scientists and engineers to invent and "get around" innumerable difficulties even prompted NASA in 1993 to decide to combine the efforts of the two countries in the development of a permanent space station. As it turned out, the involvement of the Russians allowed development costs to be trimmed dramatically. Of course, the reason was not that Russians are smarter or more economical. They are no more or less smart than inventors or engineers in any other country. They are just not spoiled by having everything available to them. Such abundance leads to mental laziness. People with fewer resources must be more inventive if they want to stay abreast of more prosperous competitors.

In talking about "get-around" inventions, I cannot help but think of jazz. This genuine American creation had innumerable fans in Russia. Dizzy Gillespe, John Coltrane, Duke Ellington, Coleman Hawkins, Louis Armstrong—these names drove the young people of the fifties and sixties wild. To us, jazz was a symbol of progress and freedom. It was the most exciting, electrifying, and desirable form of music. And the most inaccessible as well. Both the Soviet propaganda machine and the KGB took strong measures to prevent jazz from being imported from the West. They saw jazz as the symbol of a decadent capitalism, capable of luring immature young people away from the ideas of "flourishing communism." For this reason, it was considered an anti-state activity.

Jazz concerts were forbidden, foreign radio broadcasts of jazz music were jammed, and, obviously, no recordings of jazz were sold in the stores. But the younger generation did whatever was necessary to hear its favorite musicians and get around the wall of silence. Since we could not buy jazz records, local brains decided to come up with a way to produce them on their own. Obviously, owning a shop or plant for manufacturing records was strictly illegal, which meant that making vinyl discs was out of the question. Thus was "jazz-on-the-bones" invented.

To make these surrogate records, it was essential find a plastic material which was both readily available and sufficiently malleable to make the groves that carried the sound patterns. Since there was no way to buy such a raw material, some substitute had to be found. The underground producers of the copy-records began buying up used X-rays from medical laboratories and hospitals. Large, thick, and relatively soft, and with a smooth and shiny surface, this celluloid film was the perfect raw plastic medium for making the records. Simple duplicating machines were constructed, with original vinyl records smuggled from the West used as masters. The machines had a very sharp stylus driven by a solenoid, which, in turn, was controlled by a normal stylus set on the master record. The sharp vibrating stylus chiseled groves onto the X-ray sheet, copying the patterns of the master record.

This grooved film was next cut out in the form of a disk which could then be placed on a turntable for listening to the music. As a result of the copying process, you would get a replica of the original jazz record, but one imprinted on a used X-ray on which could still be seen, perhaps, a human spine, ribs, limbs, or a scull. Thus the records were dubbed "jazz-on-the-bones." Sold on the black market, they quickly became very popular. Indeed, they were so popular that the government, realizing it could never control it, ended up pretending that jazz did not exist.

Of course, celluloid being such a soft plastic, these records were not particularly durable, and after several play-backs the sound quality deteriorated substantially. To prolong the life of the "jazz-on-the-bones" records, people used wooden styluses made out of matches. Though these "needles" were good for only one play, they did not damage the records. Whenever you wanted to hear a favorite group, you needed a match and a knife. The end of the match had to be sharpened make a wooden thorn, then inserted into the phonograph cartridge and set onto the "bony" record. But listening to the music you could never tell the difference between the original and an X-ray disk.

To my mind, "jazz-on-the-bones" is a perfect example of a "get-around" invention. It might not have been terribly elegant in technological terms, and it would hardly have been of much use in any other country, but it certainly served the purpose.

I myself can take credit for a similar sort of inventive solution. My infrared ear thermometer, which I came up with in 1982 in America, was also a "get-around" invention. I knew that anyone wanting to design a noncontact thermometer that operated at room temperature had to use a so-called thermopile, since that was the only such sensor available. But thermopiles were quite expensive, and since I wanted to make a medical infrared thermometer for the consumer market, there was no way I could use something so costly. The average consumer cannot afford to spend $500 on a thermometer, so I used the "poor man's" approach. I did what I would do if I were in Russia—I began looking around for alternative ways of making an infrared thermometer, without a thermopile. I came up with the idea of using what is called a pyroelectric sensor, which was a tenth the cost of a thermopile. Nobody had ever thought of using a pyroelectric sensor for measuring temperature before, because it was never intended for that. But I invented a way to make it work even better than a thermopile, and it was a great success. It was as if I had discovered a way to play music with a hammer and the music turned out to be quite lovely.

It would be wrong, of course, to think that all we did was look for ways around problems created by people, rather than Mother Nature. Ordinarily, inventors do not worry about whether the cause of a problem is a shortage of available components or a deficiency in God's creation. They just keep forging ahead, assuming that our world is somehow unfinished and that it is the job of the inventors to pick up where God left off.

In the Soviet Union, though, there was no such thing as an inventor obtaining a real patent. Everything belonged to the state, including the citizens and the fruits of their creativity. Therefore, an inventor was granted not a patent as we think of it here, but rather a "certificate of invention," which gave him a certain moral satisfaction but no legal rights. Still, no matter what kind of a paper you got for it, an invention was still an invention.

I came up with quite a few inventions in the field of medical instrumentation. One of my favorite areas was acupuncture—or rather, an electrical variant which we called electro-puncture. I was able to find a couple of books in the library on Chinese folk medicine and studied them avidly. The books published in Russian on that subject were very good. In fact, they were far more up-to-date than anything I saw published in English. The reason for this was that most of the English-language books on Chinese medicine were written by Chinese doctors, whose views on acupuncture were mainly based on the ancient traditional approach. The Russian books, on the other hand, tended to be written by top doctors with backgrounds in the Western-style medicine, but who had studied the folk medicine of China. They were thus able to combine and synthesize the best from both the Western and the Oriental approaches, which made everything far more interesting.

I studied both Chinese acupuncture and its French variant, which is virtually the same except that the needles are inserted only into the helix and lobe of the ear. My approach, however, was not to use needles at all, but rather to send minute electrical currents through a small gold-plated contact the size of a matchstick, which we called an electrode. I designed a special instrument that was capable of locating the exact position of the acupuncture point, and then sending the electrical current to that point. As for finding the approximate location of the point and relating it to an internal organ or illness, we used traditional Chinese acupuncture maps for the body, or French maps for the ear. The effect of our needleless electro-puncture was often quite impressive. We could use it to perform both diagnosis and treatment.

One day, I was investigating sensitive points on the ear of my laboratory technician. The instrument indicated that the point which, according to the map, was related to his stomach, was unusually active. He even felt a slight pain when I touched that point. I told him that he probably had some problem with his digestive system. He said no, he felt fine. Two days later, however, he was taken to the hospital with acute stomach pain. The analysis showed that he had an internal infection.

Making an Inventor

At the research institute we used my electro-puncture device to treat migraines, neuralgia, backache, hemorrhoids, skin diseases, and many other ailments. We could even induce a period in a woman who had missed it due to pregnancy.

The acupuncture points on the human body fascinated me. I wanted to get a clearer idea of what they were and how they worked. One of the tools I used was Kirlian photography. This method involved the use of a high-voltage generator attached to a layer of salty water covered by a glass plate. The patient placed his hand on the plate and, with the room darkened, we could then photograph wonderful pictures of an aura around the hand much like a solar corona. Amazingly, the acupuncture points of the hand looked like tiny flashlight beams. To understand this method better, I traveled to the city of Krasnodar, where Semyon Kirlian lived. Kirlian had invented his method in the forties, but the Soviet government viewed it as "idealistic" and treated the inventor very badly. When I arrived at the home where Kirlian lived in poverty with his wife, the old sick man refused even to let me in. He thought I was yet another government agent come to harass him.

When I left the Soviet Union, I managed to take my electro-puncture device with me. While living in Cleveland, my friend Dr. Simon Gelman and I conducted experiments with it to induce anesthesia in patients undergoing surgery. Simon was an anesthesiologist and very interested in finding new ways to reduce pain. Unfortunately, we only worked together a short time before Simon left Cleveland to take a position in Birmingham, Alabama. At one point I visited him there and we continued our experiments briefly, but then I was offered a job in industry and our collaboration came to an end.

When I was a student at the Polytechnic Institute, one of my friends told me about the chain of underground ice caves at Kungur in the southern Urals. A group of students from the institute was planning an excursion there to explore the possibility of using the caves as natural halls in which to perform so-called color music. The institute's radio-electronic department had a design center where several students were working on various types of music synthesizers. A group of us had gotten together to develop an instrument for performing music with colors—that is, to create a light show with colored lights artistically linked to the music. In effect, they were hoping to create a new art form. This color music, as we dubbed it, was to be a high-tech version of a ballet, where the dancers were not men and women but abstract shapes which would move and develop and be transformed in accordance with the sounds of the music. It was to be a synergetic art. We were not alone in our efforts. A few other groups in the Soviet Union were trying to do similar things. One strong group was in the city of Kazan, while another was in Moscow and also in Kharkov, in Ukraine.

The idea of combining light and music was not really new, of course. Fireworks have been accompanied by orchestras for centuries. The Russian composer Alexander Scriabin wrote a symphonic poem of fire, "Prometheus," and in 1908 built a special organ with several colored electric lights to perform the piece. In 1922 the Danish composer Thomas Wilfred promoted a similar machine, which he called a "Clavilux," though these light organs remained curiosities more than anything else, for the technology was still too primitive. *Son et lumière* shows were and still are performed for tourists at the Egyptian pyramids[1]. Walt Disney created a beautiful and now classic display of color music in his brilliant movie "Fantasia," with Leopold Stokovsky conducting. Other innovative film makers, such as the Canadian Norman McLaren, have also combined the movement of abstract figures and music. We were intrigued by these experiments and wanted to do something similar inside the weirdly shaped Kungur ice caves.

We took a train to Kungur, then headed straight from the station to the snow-covered hills and the caves deep within their bowels. The caves were closed to the public, but for a small bribe the drunken guard agreed to let us in for a few hours. The caves extended for miles. Electric illumination had been installed in certain areas, and some local creative minds had

---

[1] The show by the same name can be seen in Versailles, France, and recently, I saw a magnificent display of color music in Barcelona, Spain, where the media was moving water of a fountain.

even put colored filters into the floodlights set behind the rocks and permafrost ice formations, giving the caves a mysterious, though static, appearance. Some of the caverns had a peaceful feel, while others were quite dramatic. We could almost physically perceive how those weird motionless webs of stone and ice could be vitalized by music and moving lights and shadows. I took a few dozen color slides so that we would later be able to select the appropriate music and design the lighting equipment[1].

Unfortunately, the enterprise went nowhere. Such a project had little if any chance for success in Soviet Russia. Obviously, the concept of color music struck the Party leaders as something decadent and without any apparent propaganda use which might then have justified the expense. By the time our request for permission to transform the caves into concert halls was turned down, however, we already done a lot of the preparation work, and had even composed some of the score. I did not want all that groundwork to be lost, so when it became clear that we would have to forget about the caves, I decided to build my own, portable color music organ. I came up with a design, and a couple of good friends of mine, Vladimir Bochkov and George Karmanov, helped me put together the actual machine. Bochkov and Karmanov both died young, and it warms my heart to remember the months of hard work these good men and I spent building my color music machine.

Today, in the nineties, rather than a color music instrument, I would have designed a graphics software program to be displayed on a large-screen projection TV. But twenty-five years ago personal computers did not yet exist, and decent large-screen TVs were unavailable. Everything had to be designed the hard way—which is to say, the hardware way. In effect, today's computer generated graphics are as much of an advancement on my projection-type instrument as my machine was on the primitive Clavilux designed almost fifty years before that. My color music machine resembled a large 40-inch TV with a flat screen of a thick milky glass. Inside were a few dozen tiny lamps, reflecting mirrors, six electric motors,

---

[1] The idea of performing music accompanied by a dynamic lighting in a cave eventually caught up. The creative Spaniards did it in the caves near the Mediterranean town of Nerja.

and several shadow masks and color filters. A 25-foot-long cable linked the machine to a keyboard which resembled a sound mixer from a recording studio, except that the levers on the "mixer" controlled not sound, but the intensity of the lights, the colors, the shapes of the figures and their rate of movement. The moving colored shadows and figures were projected onto the screen by the method known as rear projection.

To create the various shapes, we had to cut out many miniature shadow masks and combine them with colored plastic film. Cutting out these tiny abstract figures was a tedious and tiresome process, so instead of scissors we decided to use a hot wire to quickly burn out holes in the thick black paper we used for the masks. The masks and the colored filters could move at various speeds, which I controlled manually. The interlacing of colored shadows and beams of light produced beautiful, breathtaking effects. Actually, the machine was something like a cross between a traditional shadow theater and a kaleidoscope, where the patterns were preprogrammed and recalled at the will of the artist.

For the first performance I selected two pieces: "Dawn on the Moscow River" by Mussorgsky, and the first movement from the Dvorak's Ninth Symphony, "From the New World." The preparations and rehearsals took nearly two months. Our first performance was held at the medical electronics laboratory of the research institute where I worked. We invited a crowd of engineers and scientists for the concert. The auditorium was packed. I gave a short speech, preparing the public for the unusual performance. Obviously, no one had seen anything like it before, and so some preparation and explanation was needed. After my talk, we turned the lights off, George Karmanov switched on the tape recorder, and I began with Mussorgsky.

The Mussorgsky piece was very cinematographic. Listening to it, you could almost feel the early morning chill and see the pink glow of the sun as it rose above the glittering water of the river. This was what the visual portion of program looked like on the screen. Partly abstract and partly realistic, it was a slow and gently unfolding introduction to this new audio-visual art of color music. The audience liked it. Encouraged,

I moved into the Dvorak. The movement I had selected was very different from the peaceful Mussorgsky music—intense and dramatic, with a strong emotional flow. The entire performance with both pieces took only twenty minutes, but I was exhausted. I never knew that moving levers on a panel could be such a demanding experience.

It was a successful first experiment. From it, I learned several things. First of all, I realized that the experience of color music was far more emotionally intense than simply hearing music. The sheer volume of visual and auditory information being forced on the brain of the spectator was so much greater that after just ten minutes of watching and listening a person felt very tired. In effect, color music was an abstract ballet. Its connection with music was natural and logical, for music itself is an abstract art. Music has almost no direct association with anything from real life. It is an aesthetic tool directly influencing human emotions. The same was true of the visual element I had added to it. Being many times more informative than sound, this visual element turned out to be a very powerful artistic vehicle. It was no surprise, therefore, that even after such a short program, the audience was nearly as exhausted as I was.

After that first experience, I held many performances at my apartment. My wife worked at the Opera and had many musician friends. A number of them liked what I was doing, but others found my concerts to be either too unusual or too artificial. One musician told me, "I understand and feel music very well and don't need your visual score to improve my digestion."

During the next few years, I participated in several conferences at which color music was performed by other groups. I met experimental musicians who were trying to compose separate original scores for both the audio and the visual portions. They had even invented a special system of notation, but what I heard and saw was neither terribly interesting nor particularly inspiring. No doubt the limitation was in the technology. Personal computers were still at least ten years away, which meant that all the graphics and light and colors had to be prepared in advance and controlled manually. Naturally, the tools were very limited, and unique to the instrument being used in the performance. As one music critic I knew at the time put it, "Composers who can make music, make music. Those who can't—make the color music." There was a lot of truth in that observation. This was why for my own performances I selected pieces from well-known classical works, to which I then added my personal interpretations.

It was fun and I enjoyed it very much. As I was not a musician, this was partly a way for me to express my inner creative impulse, and partly a way to explore human audio-visual perception and emotions, a topic which interested me in terms of my professional work. I regarded these exercises as playful experiments, nothing more and nothing less, and I had no intention of ever becoming a professional performer of that art.

When the time came for me to part with Russia, I had to leave my color music machine behind. Soviet customs would never have let me take it with me. Besides, I was leaving the country in a rush, with just a few suitcases. A bulky rear-projection machine would have been far too much of a burden. I ended up giving it to my friend Vladimir Bochkov, since he was the one who had first introduced me to the idea of combining music and color and had helped me build the machine. When I had been in America for about a year, I received a letter from him in which he told me that he tried to carry on my concerts, but that the momentum was not there any more. After this our correspondence lapsed. It was not safe for him to send letters to America, or to receive letters from me. Ten years later, I learned that Bochkov had died. I have no idea what happened to the machine after his death.

Another friend of mine, Lew Rubin, was both a fine electrical engineer and an accomplished pianist. He had even won a competition for young musicians at which he received his trophy from the hands of the great composer Aram Khachaturian. I was the science advisor for his Ph.D. thesis, and we worked together on many projects—but most of all on the electronic music instruments. Lew later organized a laboratory of computer music at the conservatory, but before this he traveled to Moscow to meet a legendary inventor whose story is, I think, appropriate to tell here.

The name of that man was Lev Sergeevich Termen, but in the West he was know as Leon Theremin, the inventor of the Thereminvox, the first electronic music instrument. He built it in 1920 and demonstrated it to Lenin. The instrument consisted of a wooden cabinet with two antennas, one to control pitch, the other for volume. To play the Thereminvox, you had to wave your hands near these antennas like a conductor, and the internal oscillators would produce a sound resembling something between a human voice and a string instrument.

Theremin came to America in 1927 and lived in New York until 1938. RCA built his Thereminvoxes for a brief time, but then discontinued them. Upon his return to the Soviet Union, he was promptly arrested. In the West it was believed that he had perished at the hands of Stalin's associates. In fact, he survived, chiefly because his creativity and knowledge of electronics, together with his experience living in America, were needed by the Soviets to develop an electronic bug for its surveillance of the American embassy. They kept Theremin in a special camp with engineering and research facilities where he worked on his new—unfortunately far from musical—devices.

While living in New York, he often met Albert Einstein, who had an interest in the connection between music and geometric figures. But this did not interest Theremin. He later said, "Einstein was a physicist and theorist, but I was not a theorist—I was an inventor—so we did not have too much in common. I had much more kinship with someone like Lenin." There is no question that Theremin had a brilliant and free-spirited mind. Yet despite of his hard life and the rare opportunity to be able compare the two worlds, he remained all his life, until his death at ninety-six, an orthodox Communist. Indeed, there is no such thing as universal wisdom: it is all too possible to be smart in one thing, and still remain stupid in everything else.

> Every child is an artist. The problem is how to remain an artist after growing up.
> - *Pablo Picasso*

My friend Vladimir Bochkov was a very interesting character. Some seven years older than I, he was always youthful, vibrant, and full of original ideas. His childlike view of life and constant readiness to leap on a bandwagon made him a favorite among the students at the university. He had an amazing ability to be surprised by anything and everything. He saw nothing self-evident and took nothing for granted. My mother once told me, "I was riding the bus. The weather was terrible, it was pouring buckets. Through the window, near the bus stop, I saw Bochkov standing under a birch tree, soaked to the skin. He was looking at a leaf on the tree and laughing like a baby. He didn't even seem to notice that it was raining." That was Bochkov, through and through. I know that at that time he was particularly preoccupied with the idea of the similarities of shapes in living organisms, and curvature of the birch leaf was probably one more thing that fit into his theory.

In the early sixties, he wore a military uniform and served at the Soviet cosmodrome at Tyura-Tam, where he worked in the avionics of manned spacecrafts. I first met him when I was a fourth-year student in the radio-electronic department, in a class on electronic pulse devices he was teaching. His lectures were rather colorless, and it was clear that teaching was just a job for him. His true passions were biology and physiology. I was interested in medicine as well, and it was this that made us closer friends.

In 1966 he told me that he believed in a profound interaction between the human senses and emotions. In other words, he believed that there was a kind of cross-communication between various perceptions, say, the sense of taste and sound, or between various colors and music. In essence, he was right, and medical science even knows of some extraordinary cases where "abnormal" people have not been able separate these feelings. To them, a sound might be sour or bitter, or a color rough or loud. I once read a study by Professor A. R. Luria of Moscow University about a man, a psychic, to whom the sound of a word had a visual shape which might be quite different from the actual meaning of the word. For instance, he complained that word *swine* was "thin, long, very sharp and shiny, like a sword, totally contradictory to a fat and dirty animal." When I say "abnormal", I mean simply "different" from the majority. Certainly Bochkov believed that everyone, even "normal" people, possesses the ability to experience feelings in this way. In practical terms, he was eager to explore the art of color music. His enthusiasm, energy, and wit were so contagious that I could scarcely resist joining him on his journeys into the strange and the unknown.

Besides color music, we worked together on several other interesting projects. Bochkov was fascinated with the origin of life and its evolution. He spent hundreds of hours in libraries reading books on a vast range of disciplines—from aviation to marine biology to human physiology. He had a solid background in physics, electronics, and other technical sciences, making his view of biology unconventional and always fresh.

We spent endless hours discussing the mysteries of the human brain or the evolution of species, and even tried to develop universal principles which could be applied to all living organisms from the simplest cells to humankind. We read about many attempts to develop such a universal theory, known as the theory of systems. We were strangers to the field, and our exercises resulted in only one published paper, of which, however, I am still proud. But before this, Bochkov went to great lengths to prove his idea that Nature applies a "special law" to all organisms.

He believed that he discovered such a law and called it the "Law of 37." Basically, his idea was that any living organism which exists in a "hostile" environment (and to living organisms, everything external is hostile) for some profound reason adapts its most vital function such that this function exhibits a maximum efficiency at close to 37% of its total span. Often this function is either the optimization of energy or the efficiency of reproduction or the rate of evolving of a useful mutation. Bochkov measured the shapes of microbes, fishes, leaves, the profiles of human organs, red blood cells, rates of metabolic processes, and the growth of ant colonies. His sharp eyes sought out anything that might fit his theory, and whenever he came across further "proof" of his theory, he laughed like a child and immediately either called me or rushed over to my laboratory or apartment to share his excitement.

Often I did share his enthusiasm, but on many occasions I fear I disappointed him. My pragmatism and more logical frame of mind clashed with his rather artistic and emotional approach. I was always trying to reason and deduce, while for him mere similarity was sufficient proof. Needless to say, most scientists refused to take him seriously, but this never bothered him. The beauty and harmony of life, which he had a rare gift for seeing, were endless sources of joy and happiness to him. Despite his engineering degree and his passion for biology, he was, I think, an artist—though a strange artist, to be sure, one who never painted a picture or shaped a sculpture. His medium was the creation of strange scientific theories, the products of his imagination rather than a reflection of reality—much like my own surrealistic paintings, some of which were inspired by him.

I remember one of research projects for which I needed to find out the precise value of the specific heat of water. I got a good handbook on chemistry and, to my surprise, found that specific heat (which is the ability of water to absorb thermal energy) is not constant, as physics and chemistry textbooks teach. It changes slightly across the temperature range of water, from melting to boiling, and reaches its extremum (that is, its minimum value) at a point 37% of the way along that melting-to-boiling spectrum—in other words, at about 37°C, or 98.6°F.

That magic number struck me—what a coincidence! Was it possible that such a fundamental property of water coincided with the normal temperature of a human body by sheer fluke, without any profound reason? Einstein once said that God is subtle, but not malicious. There had to be a good reason for this amazing coincidence. Needless to say, when I mentioned my discovery to Bochkov, he was pleased, but not at all surprised. "Of course," he said, laughing, "it is the same thing—that 37 popping up again."

Of course, Bochkov was not so naive as to believe in magic numbers. At the heart of his "discovery" was not just a number, but a function which in the theory of information is know as entropy. This function, if plotted on a scale from 0 to 1, reaches its maximum at a point of 0.3679..., or approximately 0.37. Just for the fun of it, I decided to see if the entropy function fit into the curve representing the specific heat of water. It fit perfectly! Bochkov was delighted, and I was confused. I was forever arguing with him, insisting that he approached science from the back door. I was always telling him that he should not just take whatever he liked and simply see if his favorite function fit it. And here I had just done it myself and got a perfect fit. Could it be that he was a genius, and I a foolish skeptic?

I wanted to comprehend the reason behind this mysterious coincidence. It took me several months of research through many books and a hundred or more scientific papers on biology, chemistry, and physics. That I was neither a biologist nor a chemist made the enterprise all that much harder. Nevertheless, I managed to glean some understanding of why the normal body temperature is so close to 37°C—this being also the temperature at which water (which makes up the bulk of our bodies) has the least ability to absorb heat. I got to the point where I was able formulate a hypothesis of the cause, but was not skilled enough to resolve the problem fully on my own. Bochkov took no part in my research, since body temperature was just one of many examples he believed fit his theory, and he had no wish to get stuck on just that one. All the same, I kept him posted nearly daily on my struggle with this mystery and the findings I had been able to dig up at the library.

In the course of my search through the literature, I found an important reference in Japanese. It was an article by Nisiama Iwao on the structure of water. I hunted about for someone who could translate the paper for me, but in the entire city I could find no one who spoke Japanese. What was I to do? I bought a huge two-volume Japanese-Russian dictionary and set about translating it myself.

The big difficulty with doing it myself was that the Japanese language is pictographic and has no alphabet. To translate the *kanji* (Chinese characters) and *katakana* and *hiragana* (Japanese characters) I had to count strokes, identify the graphics, then look up the word by section according to the number and position of the strokes. It was quite an experience. Nevertheless, after several weeks of hard work I managed to translate two pages. The paper led me to other works on water crystals and their relationships to the molecular structures of proteins. In particular, I learned a great deal from the writings of the famous American chemist Linus Pauling.

What I discovered can rephrased in laymen's terms as follows. Specific heat is a manifestation of structural changes in a material. If the specific heat is constant, the material is homogeneous and stable. Prior to my research, I thought that water was just a mixture of H_2O molecules. However, small variations in the specific heat told me that water was not that simple, and that there were probably more complex molecular arrangements involved than simply ions of oxygen and hydrogen. From the literature, I learned that when water melts, not all the ice crystals break down. Some survive. The number of surviving ice crystals at any one temperature can differ depending on the previous history of the water. That is, water that is boiled and then cooled contains far fewer crystals than water that was first frozen and then warmed to that same temperature. I read a number of papers suggesting that water melted from ice was far more biologically active than that which is boiled. Some papers speculated that birds migrate north to nest because the melting spring water is particularly beneficial for the hatchlings. Could this also be the reason why Americans always drink soda with ice, something the people of other nations never do?

Float in their mass of H_2O like shipwrecks in the ocean, water crystals are not uniform. Their shapes are dependent on temperature. From the temperature at which ice melts (0°C) to that at which water boils (100°C), there may be five temperature zones with five different crystalline structures. Interestingly, the temperature span comprising the third zone (30-45°) is about the same as that within which warm-blooded animals can survive, while the approximate midpoint of this zone—about 37°C—is the temperature set by the body's thermal control center in the small gland called the hypothalamus, located beneath the cerebral cortex.

At first, I thought this could not be right. Surely some animals had temperatures much higher than 37°C—birds, for instance. To see whether this was in fact true, I called a researcher in Estonia whom I had met a few times at scientific meetings. While I conducted research in my laboratory using bio-telemetry systems on humans, Peet Horma performed similar experiments with birds in Tallinn. A bio-telemetry system contains a tiny radio-transmitter which relays information concerning EKG, temperature and other vital signs from a free-moving body. Horma told me that his experiments showed that the body temperature of a bird in flight was indeed very high. However, most of the time, when the bird was not active, its temperature dropped to the optimum level of 37°C.

I had to conclude that the "set" temperature of the body was governed by the structure of water crystals. Having developed this idea as far as I could, I decided to publish my hypothesis in a respectable scientific journal with the hope that someone might find it to hold water, so to speak. I wrote a detailed, strongly reasoned paper, and invited Bochkov to be my co-author, since it had been his ideas which first prompted me to undertake the research.

I sent the manuscript to several journals, all of whom rejected it as being too speculative and unconventional. Disappointed, I did not know what to do next. Then one of my associates suggested that I send the manuscript to Professor S. Schwartz, who was the chief editor of the prestigious magazine *Ecology* published by the Soviet Academy of Sciences. Dr. Schwartz was known for his open mind and his willingness to take risks. Having nothing to lose but without much hope, I sent him my manuscript.

A week or so later, Dr. Schwartz called me and asked to come by his office at the Institute of Ecology. To my delight, when we met, he praised my work and told me that he liked the paper very much. "I understand why it was turned down," he said. "People of science don't like unorthodox ideas from someone who has no name in the field. But don't you worry about that. Your paper needs some minor revisions. I want to work on it for a few days. Let's set up another meeting for next Friday, here in my office. Don't worry, young man, we'll publish it, and you'll be proud of it."

The following Friday, I showed up again the modern building of the Institute of Ecology and took the stairs to the second floor where Dr. Schwartz's office was located. As I went, I noticed several people rushing about with grim faces, and felt a tense atmosphere. I knocked on the door and saw the professor's secretary. The old woman was crying.

"May I see the professor?" I asked, having no idea what was going on.

"I'm sorry," she sobbed. "Academician Schwartz died last night."

It was tragic news for all those people whose work and life depended on that energetic man. And it was a death sentence for my manuscript.

A few months later, someone else suggested that I send the paper to a popular journal, something along the lines of the American *Popular Science*. It was worth a shot, so I simplified the article and sent it to the magazine *Chemistry and Life*, which, after some delay, published it in its September 1976 issue. The editors revised my writing so much that I hardly recognized my own work. Nevertheless, the paper was published, and I even received a good deal of encouraging mail from people who found my hypothesis interesting. Unfortunately, I never returned to this problem and still do not know whether I was right or wrong, or whether water structure indeed determines the set temperature of the human body.

5
Escape from the P.O. Box

*U*ntil the mid fifties, the vast KGB network of the infamous "Gulag Archipelago," linking thousands of concentration camps throughout the largest country in the world, existed for the purposes of providing slave labor and exterminating prisoners. But besides that terrifying web of hunger, torture, and death, the Soviet Union maintained a very different archipelago. Unlike the Gulag, the islands of this other archipelago could be called islands of prosperity. They were modernday citadels whose residents lived in comfortable and spacious apartments, ate decent food, and had access to a variety of Western-made clothes, appliances, and other sought-after goods. The inhabitants of these islands of prosperity received better medical care than ordinary Soviet citizens. Their children went to better schools and sport clubs. They even had their own private resorts along the Black Sea. Nevertheless, these people too were the virtual slaves of the Soviet state. They lived in their gilded cages by virtue of the useful talents they possessed. They were the engineers, the scientists, the technicians, and the skilled workers whose job it was to create and produce Soviet nuclear missiles and other advanced weaponry. It was in one of these citadels—or, as the inhabitants called them, "establishments"—that the famous physicist and Nobel Peace Prize winner Andrei Sakharov led the team of top physicists who developed the Soviet version of the H-bomb.

Outside the Soviet Union, only the intelligence services of the free world countries knew anything about existence of these secluded enclaves. To ordinary Soviet citizens, these places were known simply by their post office box numbers, and thus commonly referred to as the "P.O. Boxes." They did not appear on any map, were never mentioned on the radio, television, or in print. It was as if they did not exist at all but were merely the product of a paranoid imagination. But they did exist, and more than that, they managed to assemble the most brilliant constellation of scientific and technical minds in the country.

Special KGB forces maintained tight security around the P.O. Boxes. Inside the barbed-wire fences were the residential sections with their food and department stores, their kindergartens, schools, and parks. The research laboratories and manufacturing facilities stood nearby. For the convenience of those who lived and worked there, business and residential areas were laid out within easy walking distance of each other. The inhabitants of a P.O. Box town needed special permission to leave its ringed perimeter. Only a few close relatives with special passes could visit them there, and then only for a few days. All contact with foreigners was strictly forbidden. Mail was censored. Relatives on the outside, when asked about an inhabitant's place of residence, typically replied: "He lives in P.O. Box number so and so." Nothing more. Some of these places were also known by the names of a nearby village or town or cities, with the attached P.O. Box number, for instance, "Chelyabinsk-40" or "Sverdlovsk-44." Those of us who lived in the Urals and Siberia were aware of these places. Some of us had relatives there who could buy various select goods in the stores on the inside, goods which were generally in such short supply everywhere else.

Upon graduation from the Polytechnic Institute, many of my classmates were assigned to live and work in such guarded settlements all around the country. They accepted the posting willingly and with pride, for it was a sign of trust and honor. But even more important, life in the P.O. Boxes was so much easier and more comfortable that many talented young people were only too happy to go there to escape the hardships of everyday Soviet existence.

I too, like my classmates, was ordered to report to a P.O. Box after graduation. The P.O. Box stood smack in the center of Sverdlovsk, my hometown, and its employees all lived in the regular residential areas of the city, rather than an isolated compound. P.O. Box No. 320 was not in the business of making nuclear bombs. Instead, it produced radar systems for ground-to-air missiles, and so was deemed to be of somewhat lesser importance. Military production in the Soviet Union was generally divided into two groups, according to its purpose: offense or defense. Offensive devices inevitably received highest priority, and those involved in their production enjoyed measurably better treatment. Defense-oriented production trailed behind. For this reason P.O. Box No. 320, lacking exalted "gilded cage" status, was located within the city limits. But all the same, getting a job there would mean passing a security check of the first degree—a mark of "trust" that would mar my life forever.

At the time of my graduation from the Polytechnic Institute, I could not even dream about one day moving to the West, let alone ending up in the United States, the "main adversary" of Soviet military jargon. Nevertheless, for some unconscious reason, I wanted to keep my record free of any secrecy clearances. Very likely my determination stemmed from a desire to keep all my options open. That I would have no options of any kind in the P.O. Box was certain. I cared little for the material comforts of life so long as I was able to do what I liked. I always wanted to maintain my independence and stand out from the Soviet crowd. That, I am convinced, was the driving force behind my efforts to get off the list of graduates who were to be assigned to the P.O. Boxes.

But though my resolve was firm, the actual chances of accomplishing my goal were almost nil. First, I went to the P.O. Box No. 320 personnel department and asked them to give me a release paper. My request was denied. P.O. Box badly needed electrical engineers, and it made no difference to them if someone didn't want the job. I wrote to the Ministry of Education, to local officials—all in vain.

Since the medical sciences still interested me very much, I thought the best occupation for me might be a research position in a university medical school or biology department. A friend of mine, Lew Gor, introduced me to Professor Vladimir Rosenblat, M.D., a department head at the Medical Research Institute. His department included a medical electronics laboratory, and Rosenblat said he would gladly hire me, but only if

I managed to get a release paper from the P.O. Box, or somehow arranged to be reassigned by the Polytechnic Institute. Both alternatives struck me as equally impossible. It was November 1967 and my graduation day was fast approaching. If I failed to get myself reassigned before the end of the year, on January 2 I would have to report to the P.O. Box and that would be that.

One frosty morning I was standing at the streetcar stop on my way to school. A middle-aged man in a black coat stood nearby, apparently waiting for the same streetcar. I had seeing him many times before and knew that he worked somewhere at the Polytechnic Institute. We had never spoken, but this particular morning he asked me where I was going. I told him that I was on my way to Institute.

"Why do you look so unhappy?" he asked then. "Your school years are the best time of your life."

"I have a problem," I said, "and a rather serious one. I've been assigned to a job at the P.O. Box, but my interest is medical electronics, not military equipment. But there doesn't seem to be any way I can change my assignment..."

He squinted at me and smiled, "Stop by my office at two this afternoon. Room 130 in the main building. Maybe we can help you somehow." At that moment, our streetcar arrived. We got on and I turned again to that strange-sounding man.

"But what are you? Some kind of Santa Claus? How can you help me?"

He smiled again. "I'm the head of the job placement office."

It was a stroke of pure luck. When I showed up at his office that afternoon he took out a big book labeled "P.O. Box No. 320," turned to the Fs, and crossed out my name. Then he took another book marked "Free Placements" and entered my name in it. That was all there was to it. I was free to take whatever job I could find!

"Good luck to you," he said simply. To my regret, I can no longer recall the name of this man with the kind heart and smiling eyes. God bless him. With a single stroke of a pen he changed my life forever. Now I know—for the best. It is the same story the world over: the most powerful aid in getting things done is whom one knows.

Such was my first escape from the P.O. Box, before I ever set foot there. The second time was not so easy...

The Soviet state owned everything in the country: its forests, its lakes, the bowels of the earth, the birds and the beasts, the air and the clouds. And it owned the people as well.

To keep these human resources reasonably useful and productive, the state had to take care of them. Every Soviet citizen received a minimum amount of money to buy food (to provide the body energy), clothes (to protect them from cold), and for various other personal possessions and entertainments. Ownership of the tools of one's trade or of land was prohibited by law. Most other essential services were paid for directly by the state, including education, medical care, and job placements. The Soviet people were relatively free in their day-to-day routine. Yet whenever the state had need of someone, personal choice was not a consideration. You belonged to the master, to the state and the Party, and only the master had the right to decide where you lived, where you worked, and sometimes whether you even lived or not.

The progress of this world is driven by men and women of vision, not by the masses. These independently minded individuals need freedom of choice. They thrive on competition, on going against the grain, making their own decisions. They reach their greater achievements by breaking rules and following their own paths. By contrast, the masses tend not to mind a master who will take care of them. Indeed, they long for a leader to guide them, to think for them, to judge them, to reward and punish them. The Germans admired their Fürer, the Japanese worshipped their Emperor, and the Russians loved their tsar. The Soviet state was a modernday version of the tsar, looking after its subjects and ruling over their lives. The system removed the necessity of thought or competition. Everything was taken care of. It was a state-wide welfare system extending to the entire population. The actual fruits of one's labor were rewarded only if they served the direct needs of the Party. For the most part, the socialist system of state welfare crushed any show of initiative and lulled the people into passivity. Men and women lost their desire to succeed and prosper, and became content merely to survive and drift with the flow. Many liked it that way, and now, after the collapse of the Soviet system, miss it very much.

I have found the American masses to be remarkably different from what I knew in Russia. Its people are immigrants or descended from immigrants. And an immigrant is never an average person. If he were, he would have stayed back home and put up with whatever hardships he might have faced there. But instead, the immigrant has severed his ties and plunged into a strange new life, often changing his name, his profession, his language, and leaving behind friends and family alike. Many of those who stayed home carried their prisons inside their heads, while it was those who emigrated who possessed the free spirits and independent minds. The newcomers to this country were pioneers and fighters. I believe that the American people, both those who are the descendants of immigrants, and those who have come more recently, are the products of a kind of selective evolution. The free spirits came to this soil and passed their genes on to subsequent generations. Just as a bird begets a bird, and a fox a fox, so does the pioneer beget a pioneer.

Every nation or ethnic group has its own share of visionaries, of entrepreneurs, but in America they are found in much higher concentrations. I believe that this amazing sociogenetic phenomenon is one of the cornerstones of America's greatness and uniqueness. For this reason I view the mentality of the American masses as radically distinct from that of many other peoples. In Russia, I saw a different crowd, sometimes dismayingly passive, sometimes malicious and merciless. And I did not want to be a part of it.

I worked at the Medical Electronics Laboratory for nine years, earning my Ph.D. degree and inventing several medical instruments, until finally I came to the inevitable conclusion that it was time for me to move on. After my first five years at the laboratory I was already beginning to sense that I had gone as far as I could go there. Not that I was looking to be promoted to a higher position or better pay. I was a researcher and an inventor, and had no desire to start climbing the bureaucratic ladder, to become, for instance, a laboratory head. For one thing, such a position would involve endless paper shuffling, something I loathed. For another, I was a Jew—a substantial stumbling block even if I wanted to move into a simple managerial position.

Still, my job had come to bore me. I needed something more exciting, new projects, new ideas, more contacts with my colleagues. But getting a new job in the Soviet Union was a tremendously difficult enterprise. Very often, it required relocating to another city or town. Yet every Soviet citizen had government permission to live in one place and one place only. To get a job you had to present proof of permission to live in that locality, and to get the permission to live there, you had to present proof that you already had a job. It was a Catch-22 situation that one could well spend a lifetime fighting with little chance of success, like a dog chasing its own tail.

When I realized that there were no jobs for me in Sverdlovsk, and that I could not move anywhere else, I became quite discouraged. Here I was, only twenty-seven years old, and what lay in store for me? Was I to spend my entire professional life till retirement at the same desk in the same laboratory, doing the same work I had long since outgrown. Some future! So I came to the inescapable conclusion that if I could not change desks, I would have to change countries.

Though theoretically possible, my actual prospects for success were slim. If I failed to win permission to emigrate, I would lose everything. First of all, I would not be able to find work, for no one would dare hire me, except perhaps as a street sweeper or some kind of an unskilled laborer. The same would apply to my wife. This was 1973, the dark period of Brezhnev's rule in the Soviet Union. The only difference between the Stalin era and the Brezhnev years was that people were not being physically exterminated during the latter period. As Nadezhda Mandelstahm, the wife of the murdered Russian poet, put it in her memoirs, "These were almost vegetarian times." Nevertheless, neither the Party nor the state had any qualms about destroying people's lives and careers, about humiliating and crushing them morally. I had no doubt that I would lose my job the moment we applied to emigrate, and so we had to save at least a little money if we were to survive.

Another point was that at this time the Soviet state allowed one to emigrate to Israel only. Since all Jews were considered to be secret enemies of Communism and the Party, it was possible to conceive how such an "enemy" might harbor a desire to join relatives in his historic homeland. The "humane" Soviet state sometimes looked favorably on such wishes, provided, of course, that the emigration of the individual in question was "in the best interests of the state." Naturally, the best interests of the individual himself were deemed inconsequential—a slave could have no other interest but to serve his master. All other ethnic groups ostensibly lived in a "worker's paradise" and even in theory could have no wish to emigrate. Thus the only way to leave the Soviet Union was to present a formal invitation from Israeli relatives. Once in a while, this absurd unwritten law was grotesquely applied to someone the government wanted to kick out of the country. I even heard about one Russian nationalist, an anti-Semite, who was given an exit visa to Israel. Of course, whenever someone was lucky enough to make it across the Soviet border, no one forced him to go to Israel or, for that matter, to any specific country. The free world still allowed free choice, even to people without citizenship.

My wife and I decided to explore our options for getting out. One problem was that we lived in the city of Sverdlovsk, in the Urals. This region was saturated with military-related industries. Missiles, conventional and nuclear weapons, radio-electronic equipment, chemical and biological weapons were all developed and produced here. The entire area was off-limits to foreigners, and the local inhabitants were very rarely allowed to travel to foreign countries. All international correspondence was censored. In effect, it was as though we lived in a huge P.O. Box, but without its attractions and privileges. Any attempt to apply for emigration from there was almost certain to fail. The authorities would view our wish as "not in the best interests of the State." Nevertheless, we decided to start working toward a way out.

We began saving money, whatever we had left after expenses. My wife Irena worked in the Opera orchestra, and once my Ph.D. degree was approved, I got a raise. After our son Roman was born in 1975, we continued to look for opportunities to get out. When the Helsinki International Accord was signed, the Soviets pretended that they had some respect for the human rights. Shortly after the signing, we heard rumors that several families from our city had been granted permission to leave. Then, the door slammed shut again.

One option we seriously considered was exchanging our residence for one in another city or town where restrictions for emigration were not as severe as in Sverdlovsk. A Soviet citizen could relocate to another city only under one of three conditions: if the government ordered him to move, if he or she married someone from another city, or if an exchange of living quarters were arranged. Such an exchange meant that an equal number of people from another location would move into our apartment, while we moved into theirs. After that, we could apply for whatever jobs might be available there. Clearly, this last option was the only one open to us. So we started looking for a family of three somewhere in the western region of the country who might be willing to relocate to Sverdlovsk. Finding such people would be a very long and difficult process. But under the circumstances we had no choice. If we arranged the swap and moved, we would then have to live in the new place for some time before applying for an exit visa. All this could take many years.

In 1976, through the jamming roar of the radio, we heard from the Voice of America that the International Belgrade Conference was approaching. The purpose of the conference was to verify that the signatory countries were complying with the Helsinki Accord. The news brought us a fresh hope. Whenever international attention was focused on human rights, the Soviets tried to present themselves in a better light. For us, this meant that the door might open again soon.

Now, the problem was how to get an invitation from Israel, for we had no relatives there. Fortunately, the Israeli government was only too willing to send out invitations from its citizens to anyone wishing to emigrate. Of course, the whole thing was handled unofficially, but everyone pretended that it was all "kosher." In my case, I knew someone in Moscow who had a friend in Israel. The Israeli, a man by the name of Mikhail Grobman, arranged for an invitation to be sent to my home address. It was a petition for me to join a "relative" named Brosh Keti. It was sheer luck that the letter from Israel was not confiscated by the censors, who usually opened any piece of mail coming from abroad. Anyway, we got the Israeli invitation and began monitoring the international situation with even greater attention.

In January of 1977 I heard rumors that about forty families from our city who had been on the waiting list for years were being granted permission to go to Israel. Even the closed city of Sverdlovsk was opening its door a crack. But for how long? We felt the time was now—otherwise we might wait years for another chance. We decided to go for it and apply right there, from Sverdlovsk. Who knew, we had been lucky before...

The first thing we did was to quit our jobs. My colleagues at the laboratory were shocked—why? I told everyone that I was moving to another city. Most of them did not believe my story, but they pretended they did. Irena did the same thing at the orchestra. To spare our relatives any humiliation, we spread the rumor that we were moving to Vilnius, the capital of Lithuania, where I had been "offered a great job in a biomedical research institute."

We really had no choice but to resign our positions before handing in our petitions. Had we not, we would have been kicked out anyway, but not before being subjected to a humiliating public meeting where our former coworkers would be required to condemn our "treacherous" behavior. Newspaper articles depicting us as ungrateful debased citizens would then appear, along with other such signs of an organized "people's displeasure." Besides saving us from considerable unpleasantness, my resignation spared my former boss and science advisor Professor Rosenblat the burden of responsibility for my actions. In the Soviet Union, the scapegoats were usually the bosses and relatives.

As soon as we became unemployed, we walked into the building of the Ministry of the Internal Affairs in Sverdlovsk and handed in our petition. The petition was accompanied by a thick pile of documents. These papers were yet another means of humiliating the applicants. They consisted of statements and forms from various people and organizations relieving us from real or made-up responsibilities. It had taken us a few months to put them all together. But on March 17, 1977, we had everything ready, filed the petition, and began our life as nonpersons.

> The world is a madhouse, so it's only right
> that it is patrolled by armed idiots.
> -Brendan Behan

For over three months we waited without a single word from the authorities. This did not really surprise us. Some people waited for years—what were a mere three months? While we waited, we discussed our strategy for the future. If rejected, we would fall back on our old plan to move to another city. If allowed to go, we would have to learn a new language. But which? Where we were leaving from was clear, but where did we want to go? My artist friend from Moscow who had arranged the Israeli invitations for us had already emigrated. He was now living in Paris and invited us to come to France. So we began studying French and, just in case, English.

I realized that if we did go, I would be saying good-bye forever to the city of my childhood and youth. There would be no coming back. I knew that I would never again see my friends or my relatives. Never again would I walk the halls of the school where I had studied. Everything would be lost forever. Somebody once said that to emigrate is to die a little.

If they let us go, I would take with me my most precious possession—my memories. Those, and maybe few photographs. Photographs! What a great way to preserve the past! And so I began taking daily walks around the city with a camera over my shoulder. I captured on film all the places I wanted to remember: the house in which we lived, my school, the boardwalk around the city pond, the old Ipatyev mansion where in 1918 the Bolsheviks killed the family of the last Russian tsar Nicholas II, the research institute where I worked, the Opera where Irena played her violin. Sometimes, I took a train and headed outside the city to photograph the Ural forests, the grassy fields, the green mountains and blue lakes.

It was June when a friend of mine, Jean, told me about a magnificent old church with a leaning bell tower. Everyone knows about the leaning tower of Pisa in Italy where Galileo Galilei carried out his famous experiments in gravitational acceleration. But who knew about the Nevjansk tower? For centuries it had stood in a small town some sixty miles north of Sverdlovsk, and no one had ever heard of it, let alone seen a picture of it. So I decided to head out there and photograph it myself. One sunny morning I packed my camera, my passport (all Soviet citizens were required to carry an internal passport), and a sandwich into my gym bag and took the bus to the railroad station. I needed a roll of film, but at that early hour all the stores were closed, so I decided to buy some film when I got to Nevjansk.

Rush-hour was over and the local train was fairly empty. I sat in a corner seat and settled in for the ninety-minute ride. The day was hot, the windows were open, and I quickly dozed off to the monotonous percussion of the wheels. Suddenly, through my dream, I heard the train conductor announce: "Nevjansk." Opening my eyes, I saw that the train was already at the station. I ran to the door and leapt off the train just as it started moving.

When I looked around, I realized my mistake. It was the wrong station—someplace called Verkh-Nejvinsk, about twenty miles too soon. The name sounded enough like Nevjansk that I had taken it for my destination. The station building was shabby, the wooden platform old and dirty. Off to the right, about half a mile away, there was a small village at the lake side, presumably Verkh-Nejvinsk itself. On the other side of the tracks I saw a small grove and, on a hill beyond, modern buildings behind a tall fence. Now I recognized the name of the village! It was the address of one of the most secret of the P.O. Boxes, where some of my former schoolmates still worked. Sometimes it was referred to as Sverdlovsk-44, and sometimes as Verkh-Nejvinsk. It was the site of Soviet's production and assembly plant for nuclear warheads. Apparently, those buildings on the hill were part of that "gilded cage." But what did I care? I had no business there—I just wanted to continue on my leaning tower.

But what was I to do for the time being? The next train was not due for another hour and a half, so I decided to wander over to the village and see if I could buy some film there. I walked down the platform to a dirt road leading toward the gray izbas—or Russian log houses. The road was winding and dusty. The bees were buzzing and the birds were chirping their happy-go-lucky songs. A young shepherd tended his goat off in a clover pasture. It was a pleasant little walk, and in ten minutes I had reached the center of the village. There I found a small variety store which to my delight sold black-and-white film. I bought a roll and headed back to the station.

At the station, I waited for another half an hour. The sun was beating down, the air was motionless and hot. My watch showed it was lunch time. I had my sandwich with me and looked around to find someplace to sit. Nothing—just the dirty wood-plank platform without a single bench or chair. But just a hundred yards away was a lovely birch grove. Surely I would be able to find a log or stump there to sit on. I ran across the tracks and walked over to the grove. It was something of a disappointment, for the grove was not much cleaner than the platform. Amid the trees and stumps lay several rusting metal boxes, piles of aluminum and brass tubes, rolls of heavy-gauge wires and other decaying junk. It was far from ideal, but I had little choice. Looking about, I found a relatively clean metal cabinet lying on its side and sat down to have my lunch. I took the sandwich from my bag, but no sooner had I opened my mouth to take the first bite than I heard a clanking sound and a sharp voice:

"Stand up! Put your hands over your head! Don't move!"

I was baffled. What was that? Whom was the voice addressing? I looked around and to my astonishment saw that I was surrounded by about a dozen soldiers with sub-machine guns pointed straight at me. What the devil was this? With several loaded guns aimed at you, however, it is not too wise to argue or ask questions. So I stood up and put my hands on the top of my head. Immediately, one of the soldiers ran over to me from behind the bushes. He grabbed my bag, opened it and looked inside. Then another man appeared from behind the trees. He was a lieutenant. I looked at his epaulets. Green. The same color as all the soldiers. The color of the KGB guards. The soldier examining my bag took out my camera and passport. He handed them to the lieutenant, who glanced at them briefly and made a sign to the other soldiers. Another one handcuffed my wrists and then searched me. They seemed relieved to find no weapons.

When this ritual was over, the lieutenant asked, "What are you doing here, in this junkyard of strategic equipment?"

"I know nothing about any equipment," I said. "I came here to eat my lunch, and if this junk is strategic or something, it shouldn't be scattered around like this, so near the station. Take these handcuffs off, I don't want to miss my train."

"Bring the truck and put him inside," the lieutenant ordered, not bothering to answer my question. They led me to the road, where a small flatbed track waited. Two soldiers grabbed my arms and lifted me like a feather up and into the flatbed. There were two rows of plank seats. They put me in the center with my back to the cab, with two soldiers on each side and a young private and the lieutenant facing me. The remaining soldiers piled into a second truck which appeared from behind the grove. The young private sitting in front of me was very nervous. He plunged the muzzle of his AK-47 into my stomach in case I tried to catch him off guard with any dirty tricks. He kept his finger on the trigger, staring at me without blinking. No doubt he believed he was guarding a very dangerous criminal.

The truck lurched along the bumpy dirt road. Since my hands were bound, I could not hold on to the bench and was constantly falling against the guards on either side of me. They had to hold their weapons and keep me in my seat at the same time. I was particularly worried about the young soldier facing me, for I was afraid he might pull the trigger by mistake and we would both be in trouble. The idea that I was in trouble already did not quite register with me.

"Listen, pal," I said, "you either take your finger off the trigger, or move that muzzle away from my belly." He did not move or even blink. However, the lieutenant motioned to him and the soldier shifted his finger slightly from the trigger. That was better. I looked down at the handcuffs and smiled. They were marked "Made in USA." Thanks, Uncle Sam!

"Well, well," I thought. "How stupid! I'm going to miss my train. Why have they arrested me? Maybe this lieutenant's just making his daily arrest quota? Obviously, they have let me go, because there's really nothing to it. But what a nuisance!"

I was not scared at all. The entire situation struck me as rather comical, and even surrealistic. Arrested in a junkyard while eating my lunch—hardly a crime, even by strict Soviet standards. True, I had a camera with me, but it was not even loaded. I hadn't picked anything up from the junkyard. Anyone looking at this situation would surely realize that I would have to be released. Such was my naive train of thought during the short truck ride.

The truck slowed and came to a stop. I turned my head—and what I saw sent a chill through me. We were sitting in front of the barbed-wire gate leading into the P.O. Box zone. This was a very bad sign. I knew they would not be taking me inside if they had the slightest doubt of my guilt. Once the truck crossed the line of the protected perimeter, I would have very little chance of ever getting back home. The P.O. Box was a top secret facility, and no one could enter without special clearance. And anyone who did manage to get in would never be free to leave. I had no security clearance and no business being there. Nevertheless, I was being taken there against my will. What a moment before had seemed an annoying detour in my outing to see the leaning bell tower, was slowly but surely turning into a one-way trip. Such a journey could very well last a lifetime. The truck entered the compound and the heavy gates clanged shut behind us.

We rode through a sort of alley between towering industrial-looking buildings. These, clearly, were the research and production facilities. It is usually difficult to guess what is produced behind the brick walls of a factory, but if you pay attention to the details, you can figure out a great deal. I noticed several transformer substations with six-kilovolt power lines entering one of the buildings. Evidently, it housed some type of power-hungry equipment. I also saw a reactor building, which I recognized by its distinctive airlocks.

The truck made frequent stops while the lieutenant ran into various administrative buildings which stood along the road. I had no idea what he was doing. Quite likely he was unsure what to do with me and needed to consult with his superiors.

Meanwhile, I looked around, taking advantage of the free, albeit unguided, tour through the heart of that top-secret circle. It is hard to explain, but my mind seemed to work like a camera, taking snapshots of everything I saw. Even now, seventeen years later, I close my eyes and see everything clearly, as if it were yesterday. I see the uranium refining facility, the loading docks and the large green boxes marked "Device type number so-and-so," the uranium storage bay, the machine shops, the electronic assembly buildings. I see the trees, bushes and grass growing on the roof tops, the small artificial lakes, the railroad tracks with the camouflage netting above them. The place was obviously designed to be quite innocuous from above. I had no doubt that high-orbiting satellites had tried to photograph this neatly hidden compound—but had anyone ever managed to take a picture from down where I was sitting handcuffed on that miserable summer day? The more I saw the more I realized that after such of a tour, I would never be released. Certainly not for many years.

I began worrying about Irena. She had no idea where I was or what was happening. She had stayed home with two-year-old Roman. I had told her simply that I was going to take some pictures and would be back in the afternoon. When would I ever see them again?

After about forty minutes spent driving back and forth through the compound, the truck entered the residential area and pulled up in front of a five-story apartment building. A sign at the entrance, however, announced the "Civil Defense Headquarters." This surprised me, for the civil defense had nothing to do either with me or with the KGB guards. The soldiers hauled me down from the flatbed and led me to an office on the second floor. It was fairly small, obviously designed as a typical three-bedroom apartment but modified for office purposes.

I was led into a room with a large desk with two telephones, a heavy-duty floor safe, two file cabinets, a bookcase and a closed-circuit TV camera in one corner of the ceiling. Two portraits hung on the wall above the desk—one of Felix Dzerzhinski, the founder of the Soviet secret police, and the other of Yury Andropov, the then KGB chairman. The portraits explained everything. I was in a KGB office and the "civil defense" sign outside was just a front. The lieutenant unlocked and removed my handcuffs and ordered me to sit on a chair in front of the desk.

I waited for a few minutes. Then the door opened and a man of about fifty-five came in and walked straight to the desk. He wore a military uniform with the green epaulets and the three stars of colonel. His hair was gray and his brow wrinkled, and the eyeglasses on his nose were the same shape and style as in the portrait of Andropov. In his hands he held my passport, my roll of film and the camera. He sat down at the desk and said,

"Let's get to know each other. I am KGB Colonel Vereschagin. Your name I can see in your passport. Is this your real name?"

"Yes, it is."

"Well, tell me, what brought you here?"

"That's a funny question," I said. "You know that your lieutenant brought me here."

"Oh, you like jokes! So do I. We're going to have a lot of fun together. Of course, you understand that I am interested in the purpose of your visit to Verkh-Nejvinsk and the strategic equipment dump."

So I told him about how I had wanted to take pictures of the leaning bell tower, and how I dozed off on the train and got off at the wrong stop, and how I had wanted to eat my sandwich and thought the grove would be a better place to eat than a dirty platform. He listened without interrupting, taking notes in his pad. An ironic smile settled on his face, and when I finished he said, "Well, that sounds like quite a fairy tale. So simple, so innocent. And so unbelievable. What did you do in the village?"

The question told me that they had spotted me long before I began eating my lunch. Apparently, his people had seen me going to the convenience store. I explained that I had needed film and had gone to the store to buy some.

"Why didn't you buy film in Sverdlovsk? What did you want to see in the village?"

"I only realized I was out of film last night, and all the stores were already closed. This morning it was too early, so I planned to buy it in Nevjansk, but since I had to wait here for the next train, I went to the village to see if I could get it there. And there's nothing to see in the village."

"Is this the film?" He showed me the film package.

"Yes. I bought it in the store."

The colonel pressed a button on the table, and the door opened and a short young man in civilian dress walked in. His cheeks were crimson and his eyes cold and colorless.

"Ivanov, take this roll of film to the village store and check to see if it was bought there."

When the young man had left, the colonel opened my passport again, then gave me a long and thoughtful look from behind his eyeglasses. Apparently, he was looking at the fifth item on the first page, where it gave the bearer's ethnic origin. In my passport this was "Jew." For any government official this was an immediate red flag, so I was not surprised when the colonel asked, "Do you have any relatives or friends in foreign countries?"

"Yes, I know one person abroad."

"Can you tell me who he is and what kind of relationship you have?"

I told him I had a friend who lived in Paris. There was no point trying to hide it since I was receiving letters from France. The colonel then spent some time circling around the subject of my friend, asking if I spoke any foreign languages, read any foreign magazines, and so on. All his questions had to do with my contacts abroad.

Soon the crimson-cheeked Ivanov reappeared. To my astonishment he said, "No, Comrade Colonel. He didn't buy any film in the village store."

"Look here, Colonel Vereschagin," I said very angrily. "Your man is lying. I don't think he even went to the village. It's easy enough to check. Every roll of film has a lot number printed on the package. Have him go there and compare the number with the stock. And if he lies again, I won't be answering any more of your questions."

Surprisingly, the colonel agreed. "Ivanov," he said, "go back to the store and check again."

Shortly after Ivanov left—this time, it seemed, really going to the village—the telephone rang.

"Yes, very good. Bring it in," answered the colonel.

The door opened and another young man in civilian clothes came in with what appeared to be a teletype printout. Vereschagin adjusted his eyeglasses and began reading. I sat there watching him. It was clear that whatever he was reading gave him great pleasure. Every time he saw something interesting, he underlined it with a pencil and peered up at me with increasing curiosity. Finally, he put the printout aside, his eyes beaming with delight and he looking very excited.

"Well, well. Do you know what this is? I'll tell you. This is a summary of your file. We just got it over the teletype from KGB headquarters in Sverdlovsk. It looks ve-e-ery interesting. I see that you speak Esperanto. Why did you learn a language nobody in this country speaks?"

"That was ten years ago. I just liked it. I haven't used it in a long time. Is it a crime to speak that language?"

"Oh, no, no, it's not a crime. Just another piece in the mosaic. I see you applied for emigration to Israel. I also see that you have a good background in physics, electronics, fine mechanics. I have here the list of your friends and relatives. You've received some letters from America as well. You didn't tell me about them. Two from Cleveland, one from Palo Alto in California, and one from Houston. And you told me that you don't know anyone in the U.S.A. How is that?"

"These are just scientists. They sent me reprints of their papers. I really don't know them."

The interview continued like this for a while, until Ivanov reappeared and reported that I had indeed bought the film at the convenience store.

"It makes no difference *where* he bought the film," the colonel said. "What's important is *what* he managed to get on that film. Ivanov, take this film to the laboratory for developing." Ivanov took the film and, to my horror, the colonel gave him my camera as well!

I tried to protest, pointing out that the film had not even been loaded in the camera. It was no use. I felt ill. It was clear that Ivanov was going to load the film into my camera and take a slew of compromising pictures to be used as crucial evidence against me. With "evidence" like that, my future looked pretty dark.

Vereschagin continued, "I admire your fortitude. All the facts point against you, and yet you sit there pretending to be innocent as a lamb. Let's see what we have so far. You were planning to leave this country forever and wanted to secure a comfortable life for yourself abroad by selling our state secrets. I don't know yet who you report to or how, but in due course we will know everything. This establishment is a top secret facility. We know that the CIA flies satellites above us day and night taking aerial photographs, but these photographs are very controversial and of little use to them. There is no doubt that the special services of the western nations are dying to get ground pictures of this place. And here you are, a 'tourist' on his way to visit a leaning tower who through some pathetic accident finds himself right at our doorstep. You have a camera and you got very close to the protected perimeter. We still have to see what you got on film, but I'm already sure there are a good many interesting shots. If I were in your shoes, I'd start talking, and the sooner the better. We can be very understanding with those who cooperate."

At first, I tried arguing with him, offering my explanations and pointing out the absurdity of his accusations. But I was soon totally exhausted and realized there was no talking to him. His mind was made up and there was no way I was going to convince him of anything.

I could understand his excitement. This old man had no doubt been sitting in that dreary outpost for years, his one remaining ambition being to retire peacefully in five years or so, and after that, to keep himself busy playing dominos with other KGB pensioners in the courtyard of his apartment complex. And here, out of the blue, he catches a spy working for CIA, or the SIS, or the Mossad, or some other foreign intelligence.

Did he really believe I was a spy? I think he believed it might just be possible and was not about to miss his chance. And to strengthen that chance, why not enhance it with some creative manipulation, like filling my film with compromising photographs? That would finish me outright and, at the same time, improve his position immensely. What luck! What a feather for his cap! They might even promote him to major-general. His retirement would then be profitable as well as honorable, for a general's pension is a good deal higher than a colonel's. Yes, Colonel Vereschagin was a very fortunate man! But one man's fortune can be another's misery—and the very miserable man sitting across him was me.

I sat there, looking listlessly out the window, giving him short mechanical answers, yes and no. Sometimes he went out of the room for five or ten minutes, leaving me alone. Then, he would come back and continue his endless questions about my life, my friends, the plans I was making for living abroad, my interest in cinematography and science, my inventions, and on and on. The interview—or interrogation—was exhausting and interminable. I began to feel like I had already started serving my time. What he was going to do with me now? I was certain they had no prison on the premises. After all, this was a high-tech facility, not a camp. Most likely, he would transfer me to KGB headquarters in Sverdlovsk, which had plenty of barred cells and an abundance of skilled interrogators.

It was nearly six o'clock and Vereschagin was no doubt ready to go home, enjoy his dinner and toast his lucky day with a glass of chilled vodka. He stepped out of the room again, then returned and sat down at his desk, looking me straight in the eyes.

"Well, what shall I do with you? I've listened to your excuses, read your dossier, and after considering all the circumstances of this situation, I have to conclude that there is nothing to it.

I think you just behaved stupidly, but with no malicious intent. Therefore, I'm letting you go, but be careful next time where you chose to travel."

I could not believe my ears. What a twist! The most unexpected turn! Maybe the colonel was human after all? If so, how had he ever managed to survive to his current rank within the sinister bowels of the KGB? Whatever the explanation, his reversal was like a blessing from heaven. The colonel pressed the button and the red-cheeked Ivanov came back in.

"Ivanov," said the colonel in a dramatic voice, "take this violator and kick him out off the grounds of this establishment. And make sure that he gets on the train and doesn't wander about again."

Ivanov grabbed me firmly by the sleeve and we went out. As I was leaving, the colonel said, "But your camera we have to keep here for the further investigation." I was so happy to be able to go that his words barely registered. What kind of investigation could there be, if everything was resolved and I was free to go home?

Ivanov and I climbed into a chauffeured jeep which drove us through the small camouflaged gate and directly to the railroad station. Several people were already on the platform, waiting for the next train to the city. I even noticed one familiar face: the young shepherd I had seen during me walk to the village to buy film for my camera. There were a few other villagers as well: a young couple, several women, and two or three young men, probably students. Ivanov handed me back the return ticket I had bought that morning in the city. When the train arrived, he waited until I got into the middle car, then turned back to his jeep. The train stood for only two minutes, then, whistling cheerfully, pulled out of the station, and I left that unfriendly place forever.

> This beast is very wicked—
> when it is attacked, it defends itself.
> —*Jean de La Fontaine*

I sat on a corner bench near the exit. The young couple who had been waiting on the platform got into the same car. They sat across the aisle and chattered away like love-birds. The train wheels played their usual percussion and for the first time since that morning I realized that I was terribly hungry. The colonel had confiscated my sandwich along with my camera. No matter, I could survive a couple of hours more until I got home to have my supper with Irena.

After about half an hour the train suddenly stopped in the middle of nowhere. I glanced out the window and saw several people running alongside the train, looking in at the windows and under the cars. Funny, but they were the same men and women I had seen on the platform as I boarded the train. The young couple in my car jumped to their feet and also looked out the window. I had no idea what was going on, but a few moments later the train started off again and everything returned to normal. Probably some technical problem with the engine, I thought. But then why were all those people running about outside? But what did I care anyway? It had nothing to do with me.

I tried to relax but could not. The excitement of the day had left me keyed up. I decided to stretch my legs a bit with a stroll through the train cars. But imagine my surprise when my innocent walk from my car to the next created a noticeable disturbance among the passengers. The young couple stood up and followed me. The shepherd in the next car also decided to stretch his legs and went to the following car. To my astonishment, some dozen or so people who had boarded the train with me in Verkh-Nejvinsk set themselves into motion. Why? The answer was all too simple. These love-birds and students, peasants and shepherd—they were all in the service of the KGB. I was not a free man, as I had thought I was. And the colonel, far from being a compassionate and understanding man, still held a tight grip around my throat.

The colonel had not released me, he had just changed the scenery while posting his costumed guards around me. What was his game? Why did he pretend I was free to go and then send all these tails after me? I went back to my seat and tried to sort it all out. The explanation was seemed obvious. The colonel suspected I had an accomplice or accomplices somewhere who were planning a rendezvous with me, perhaps onboard the train, perhaps at one of the stops, or else in the city. Clearly, if nothing took place on the way to Sverdlovsk, the KGB would continue the surveillance there. There would be other tails from KGB headquarters. There was no doubt in my mind that they would re-arrest me as soon as they realized that I had no intention of meeting anyone. The situation was hardly any better than an hour earlier. I was still the colonel's captive.

But wait a minute! How had I been able to spot all these con-artists? Wasn't surveillance supposed to go undetected by the subject, that is, by me? And yet I had picked them out—and I was certainly no spy. Whatever I knew about intelligence and the like came from movies and books. I had absolutely no experience in such business, still I had spotted the tails within a few moments. If I was no good at this spy stuff, they were even worse. Far worse, in fact! They were a bunch of amateurs. Presumably they were just local KGB informants with no real experience in operative work. But they were all the colonel had. Obviously, I would not be so lucky when the train arrived in the city. I had heard it said that the central operatives were real pros who made their living by shadowing people day and night. I might be able to outsmart these amateurs, but I would stand little chance against the real specialists. Things continued to look grim.

The train was approaching the city. More and more people were boarding at every stop. Soon the car was packed. The closer we got to the last stop, the tighter my tails clustered about me. I decided to play it cool and show no sign that I had figured out their true identity. Let them think me a babe in the woods. I knew, of course, that as soon as we arrived these amateurs would be replaced by other tails, though it really made no difference who was shadowing me. There was nowhere to hide.

Finally, the train pulled into the city terminal and the crowds of passengers poured out onto the platform. I got out with them and headed toward the exit. My guardian angels followed me at a close distance. They were obviously nervous and afraid to let me out of their direct sight. Some of them were so close that I could even see them talking to one other, whispering into microphones hidden in their collars. This is why that the people of that brave profession are nicknamed "whisperers."

I decided not to take the bus and not to go home, at least for a while. Instead, I walked to the station plaza, crossed it, and headed right, toward a side street. I needed time to think about my situation and calculate my next move.

First, I decided not to try to make any phone calls. My going home or calling someone would trigger my immediate rearrest, and that would be that. If I went home, there was no way to continue the close surveillance, and the "whisperers" might become more aggressive. Calling from a pay phone was also dangerous. The call would be almost impossible to trace, and so would mean they were no longer in control—and the KGB did not like losing control. So I walked slowly along the streets, trying to figure out how to check behind me. I was dying to see who was following me now. I was sure I already had new tails. Would I be able to recognize any of them?

The evening was still light and the weather warm and pleasant. Summer nights in the Urals are very short, and dusk does not come until nearly eleven. There was a bakery in the middle of the next block. I was hungry as a bear. Stopping there would be a fairly logical move. I went into the shop and began choosing among the pastries, at the same time checking out the reflection in the display case.

What I saw surprised me tremendously. I recognized one of the "peasants" from the train staring in at the window. I bought a roll and left the shop, eating as I walked. I spotted two more familiar figures—one across the street and the other just turning the corner. What a puzzle! Why had there been no change of actors, why were the same provincial tails following me here, in totally alien territory? I was surprised that the local KGB would allow a team from the provinces to follow me about the city, but who could fathom their logic? Or could it be that these people were acting on their own and that the Sverdlovsk KGB knew nothing of what was going on here? To find out, I kept walking. If my followers were in contact with the local KGB office, they would be assisted by at least one car, and possibly two.

I knew the area very well. I had grown up near the railroad station and was familiar with every house and every corner. I headed for the next intersection and turned down a one-way street, then turned again onto another one-way street. If they had a car somewhere beyond those buildings, it would have to start circling the block and I would see it at one of the intersections. I wove in and out like that for some time but did never saw a single car, only my tails following me like shadows.

When I realized that the local KGB was not involved, my spirits picked up. I guessed that Colonel Vereschagin was still running things. More than likely, he had decided to hold off making a report of the day's events to his superiors and to handle the job himself. Maybe he thought that if he handed me over too soon, all the laurels and glory might slip through his fingers to the city KGB and he would get nothing for his efforts. He wanted to do as much of the work as possible. Arresting me at the P.O. Box was not a big deal, but discovering my connections and leads would be a coup that could seal his victory.

Apparently, the fact that I was not going home, but continued wandering about the city, had put Vereschagin and his people on the alert. I had no clue where the colonel was at that moment—perhaps just around the corner, perhaps sitting on a bench at the railroad station with his radio-monitoring equipment, listening to the whisperings of my followers. He was playing cat-and-mouse with me. A couple of years before I had seen the American cartoon "Tom and Jerry" on TV. Okay, Colonel Vereschagin, I thought, if you want me to be the mouse, I'll play Tom and Jerry with you. And I'll be a terrific Jerry.

My plan was this. The colonel and his people believed that I was an inexperienced amateur spy on his way to meet an accomplice and report on what had happened that day and what had I seen at the P.O. Box. I decided to reinforce that belief. Moreover, I would create the illusion that I was no mere amateur, but rather a well-trained professional spy. This would make me seem even more valuable than before. But it was also, for the colonel, a double-edged sword, since defeat at the hands of a professional spy would be that much more disastrous for him. I had seen his people in action and knew how incompetent they were and that I was smarter than they. Besides, my followers could not possibly know the city as well as I did. I had the advantage in brain power, while they had the advantage in manpower. We would see which was more efficient.

I needed to act swiftly and surely, for the slightest mistake could be fatal. I had to shake them off in a totally unexpected manner—they had to be caught off guard. Human psychology being what it is, people tend not to say, "Oh, I lost because I was no good." Instead, they say, "Oh, I lost because I was playing against someone who was very good." So, I had to make sure my followers would report to Vereschagin, "We lost him, because he was a pro." My reasoning seemed logical enough, but was Vereschagin a logical man?

Shaking them off was a piece of cake. I headed for a street where I knew there was a bus stop. I paced myself so that I would be able see the bus coming in a gap between two buildings at the end of the street. When it reached the stop around the next corner, I would be there just in time. My scheme worked. I was walking normally with my followers some distance back, but then, turning the corner, I suddenly leapt into the bus through the rear door. Most of the whisperers were too far away to do anything. But one of them who was closer ran like hell and stepped in front of the bus as it was starting to pull away from the curb. He held up his red KGB ID card and the driver stopped at once. The man got on through the front door and stood there watching me intently. I was sitting at the rear, looking out the righthand window as if nothing had happened. The bus started moving again. I turned my head to make sure there was no car behind us. There was not. Now, we were one against one.

We rode for about fifteen or twenty minutes. I pretended to be dozing off and dropped my head to my shoulders. All the time, of course, I was watching the man at the front of the bus through half-closed eyes. Before long, he relaxed a bit and pretended he was just a regular passenger, just as I did.

As we approached one of the stops, I noticed through my lashes a taxicab unloading its passengers. The bus doors opened, but I went on "sleeping" peacefully. Not until the bus was starting to move and its rear door was closing did I leap to my feet, slipping through the door and jumping into the taxi. Giving the driver all the money I had, I said, "Drive like hell." I gave him a made-up address on the opposite side of the city from my home. I looked back and saw the whisperer making a scene in the bus. The bus stopped again and the unfortunate man-hunter jumped out and stood there in the middle of the street, desperately watching the taillights of my taxi and trying to flag down a passing car. But I was already long gone.

I got out of the taxi in a deserted neighborhood and walked about again for another hour or so to make sure nobody was following me. I was alone. It occurred to me that Vereschagin had probably dispatched his people to my home at once to see if I had gone there. And when they told him that I had never shown up, he would realize that he had lost. What was I doing during those several hours, where had I gone, to whom had I talked? He had no clue—and for this reason he had lost.

After what I had seen at the P.O. Box, my escape became a time bomb for him. Now I was the one in control and he was the one in the miserable position. It was too late to inform headquarters. They would never forgive his stupid mistakes. His carrier would be ruined. He could not arrest me that night, or the next day, or ever, because he would have to hand me over to his superiors, and once I told the interrogators how I had slipped through his fingers, he would lose everything. The KGB did not look favorably on dishonesty in its own people. His position was not good at all: he could not afford to arrest me and he could not afford to let me remain at liberty. If I talked—to my wife, to friends, to relatives—sooner or later the KGB would find out, for it had eyes and ears everywhere.

And what about his own whisperers and the crimson-cheeked Ivanov? Actually, Vereschagin was probably less worried about their talking. He could always come up with some kind of explanation, perhaps telling them the whole thing was a training exercise that they had failed miserably. People tend not to brag about their own failures. But I was out of his control and that was a real threat for him. It would be essential for him to silence me. I began to fear that his only solution might be to have me killed. But even for a KGB colonel this was no simple matter and indeed might prove riskier than leaving me be. If I were killed, there almost certainly be an investigation, and who knew where it could lead? All in all, his position was very shaky indeed, and I had no idea what he would do about it.

It was midnight when I finally went home. My hands were shaking as I opened the door. What if they were there? Irena and Roman were sleeping peacefully, but that night I could not close my eyes. I had played the game the best I knew how, and thought I had predicted all the colonel's moves as well. I had won this round, but what was next? I could think of no reasonable way out for him. In the morning, I kept myself busy, making phone calls and visiting my parents. I told my friend Jean that I had gone to Verkh-Nejvinsk to photograph the leaning tower, but that my camera was stolen on the train.

That evening, I went out to the bakery, which was half a mile from my home. I bought some bread and was walking out the door when someone tapped me on the shoulder. I turned around. At first I failed to recognize him in civilian clothes. It was Colonel Vereschagin. He was wearing a brown raincoat. His face was ashen, and he looked very tired. It seemed he had not sleep the night before either.

"Let's talk very briefly and quietly," he said. "Everything that happened yesterday—you must forget forever. If you mention it to anyone at all, I'll kill you and that person, too. It would be best for you if you moved away from here—immediately."

"How I can move, and where?" I asked.

"You are waiting for permission to go. I have connections—you know where. I may be able to help you get out. But remember, keep your mouth shut. Agreed?"

"You can count on me," I said, not yet believing what I was hearing.

"By the way," he said, "here's your camera. You can keep it." The colonel turned and walked to the bus stop, and I went home to my family.

Four days later I received a telephone call from the Office of Internal Affairs. The caller told me to report there with my wife the next morning at ten sharp. Though I knew what

Family photograph filed with petition to emigrate

was coming, I still could scarcely believe it. The following morning we left Roman with his grandmother and went to the office. After waiting half an hour, we were ushered into a small room where a middle-aged woman in a blue militia uniform sat. She glanced up at us with emphatic disgust.

"You've applied to join your aunt in Israel," she said. "Our government considers family unification an act of humanity. Comrade Yeltsin has decided to grant your request and allow you to emigrate to the State of Israel. Do you accept this decision?" At that time, Yeltsin was the Sverdlovsk city boss. In fact, all decisions to let people emigrate were made by the KGB, but for form's sake the city tsar had to rubber-stamp them.

"Yes, we do," we said.

"Well then, we grant you this permission on one condition. You must leave the Soviet Union within three days. Also, you must surrender all your diplomas, credentials, and all other documents you may have in your possession. Furthermore, by the order of the Supreme Soviet of 1948, every person emigrating to the State of Israel automatically loses his Soviet citizenship. Do you understand that from now on you will be stateless?"

It goes without saying that this was a "loss" we were only to happy to incur. Three days, however, seemed unrealistic.

"It's impossible to leave the country within three days," I protested. "We need to go to Moscow, and that alone will take at least two days. Then, we have to get the Israeli visas through the Dutch embassy and then transit visas from the Austrian embassy, and after all that we have to buy the plane tickets to Vienna. We need at least a week. Otherwise, I can't accept your permission."

Naturally, I was bluffing—we would accept no matter what—but three days was really too short a time. She said it was not her decision and that she needed to consult with her superiors. She went out of the room, then came back and said, "You have ten days. And if you stay a minute longer, you will be arrested for violation of the passport regulations."

Two days later we embraced our parents and friends at the railroad station and departed for Moscow. Obtaining the visas was relatively simple, and the Dutch and Austrians were very helpful and understanding. Getting the airline tickets to Vienna, however, was a problem. We had only three days left before the expiration of our exit visas, but the clerk at the Aeroflot ticket counter said with a chilling smile, "Everything to Vienna is sold out for at least two weeks."

Obviously, that was not true, but what could we do? Then she added, "We do have a few first-class tickets left, however. They go for double the price if you wish to buy those."

Double, triple, whatever— anything to get out of there! We gladly paid the price, got our tickets, and just one day before the deadline were at the Moscow International Vnukovo airport. Our luggage consisted of just four suitcases filled with books, record albums, Irena's violin, and some personal clothes. We hugged our mothers who had come to Moscow to say good-bye, as they thought then, forever. Roman was excited about flying in the big airplane and could not understand why his grandmas were crying.

In the cabin, we were greeted warmly by the crew, who took us for high-ranking Soviet officials. Why on earth would anyone else be flying first class when the rest of the plane was nearly empty? The service was good, but we gave little thought to it. As the plane took off, we looked down on the land where we had been born, raised, and educated. The land where our closest relatives and friends remained. The land we had once loved and learned to hate, and which we would probably never see again.

Two hours later we landed in Vienna airport Wien-Schwechat.

6
Spooky Business

> People used to go off to war, but modern science can now bring it to your doorstep.

When Irena, Roman and I arrived in Austria, we expressed our wish to settle in Paris. We had there a friend, an artist, who had suggested that we come to France. I called him from Vienna and asked him to arrange French visas for us. We knew that this was going to be a very difficult process which might take several months, since France issued visas to stateless individuals only reluctantly. We wanted to remain in Austria for however long it took to obtain the appropriate French documents. For the interim, the Tolstoy Foundation, a charity organization, saw to our welfare. It helped us to buy food and some essential clothing, and to pay for temporary lodging. The Foundation was receiving its funding from royalties bequeathed by the great Russian writer Count Leo Tolstoy, the author of *War and Peace*. His daughter Alexandra, who at that time lived in New Jersey, administrated the fund. So it was that for our first few months in the free world we were virtually supported by Count Tolstoy.

In Vienna, we visited Mr. Rogoyski, the director of the Austrian branch of the Foundation. This amazing gentlemen was fluent in nearly twenty languages. Usually, he and I spoke French together, but I also heard him speak with Koreans and Bulgarians in their own tongues. The Tolstoy Foundation offered assistance without regard to ethnic origin or religious beliefs. The refugees in Rogoyski's office flocked there from all walks of life and from every Communist country: Russians, Chinese, North Koreans, Poles, Lithuanians, Czechs and Slovaks. With the help of the Foundation, we rented an apartment at Rossaurgasse 3, not far from the Ring, the central circular boulevard of Vienna.

Virtually every refugee from the former Soviet Union attracted the attention of the Western intelligence agencies, whose job it was to monitor everything that went on behind the Iron Curtain. Naturally, the majority of these refugees knew few if any secrets. Nevertheless, each of them carried some minute grain of information which could be useful for filling in the ever-changing mosaic of the Soviet Union and its satellite nations. It was a painstaking and often tedious job for many agents, but it had to be done and it was done.

The newcomers to the West arrived from the Soviet Union through two major European capitals: Vienna and Rome. But their first stop was always Vienna. Those who planned to remain in Europe stayed in Vienna, while those planning to go overseas were within a few days relocated to Rome. The refugees often had to spend several months in these cities before resettlement could be arranged. During this period, they were screened by the counter-intelligence agencies of the Western countries to fish out the spies that the KGB occasionally dropped into the human river flowing from the East. I had heard about this process from other refugees, and so was not surprised when one day I received a telephone call from the American embassy and a man speaking Russian asked me if I could come over to the embassy to answer few questions.

The American embassy in Vienna was housed in a beautiful building with a wide marble staircase in its lobby. I identified myself to a marine guard who picked up a phone and spoke to someone upstairs. A moment later I saw a short man walking down toward me. He greeted me in broken Russian and led me to a small room on the second floor. Introducing himself as an FBI agent, he said that he had a number of questions he would like me to answer.

We talked for several hours. It was a thorough and quite exhausting interview, more of an interrogation than a conversation. Indeed, it bore a striking resemblance to my grilling by

Colonel Vereshchagin just a few weeks before. The FBI agent asked virtually the same questions as his Soviet counterpart and almost in the same order. He wanted to know everything about myself, my parents, my friends, and my plans in the West.

There was, however, one very significant difference between the two sessions. During Vereshchagin's interrogation, I had been depressed, for, with every new question about my past, my future had loomed before me as a murky abyss, while in Vienna I saw everything in bright and cheerful colors. I understood the reason for the meeting with the FBI agent and did my best to help the man, and thus to help myself. Toward the end of the day, the agent told me that he was fully satisfied with my answers and felt comfortable having me talk to some other people. Specifically, he said that an agent of the CIA would like to meet with me the very next day.

The following day I was met at the embassy by a middle-aged man who showed me his CIA identification card and said that I had been cleared by the FBI. Now—if, of course, I had no objections—he would like to talk to me about Soviet science and technology and any facilities in the Soviet Union I might have seen. I replied that I would gladly share with him whatever I knew. Indeed, I felt a moral responsibility to be helpful to the United States in any way I could. Even if I was not planning at that time to live in America, I firmly believed that it was the only world power that was both willing and able to keep Communism at bay. I was only to happy to be of some use and tried to recollect any details that might interest the CIA agent. We talked for many hours—first in Russian, than switching to English, as my English was somewhat better than his Russian.

The agent brought out a large satellite map of Sverdlovsk and asked me to make corrections to it. The Soviets never published city maps, and even in official atlases the coordinates of all cities, towns, rivers, and lakes were deliberately shifted about at random to confuse foreign enemies. I was astonished to see that the CIA map contained so many very old street names and that a great many buildings were incorrectly marked. Some of the names had not been in use for thirty years or more. The agent explained to me that the CIA had few opportunities to learn about such changes and to correct the map. This was why they needed my input.

When I began to tell him about my arrest in Verkh-Nejvinsk and my unexpected tour through the P.O. Box territory, the agent said, "This sounds too important for me just to record it. You should tell this story to our people who monitor these places. Will you excuse me, I have to make a telephone call."

He left the room, returning about ten minutes later. "Our man in charge of operations against the Soviet Union is flying in today from Munich to see you. He is stationed at our European headquarters there and is very interested in your story. Meanwhile, we would like to hire you for few days as a CIA consultant. We'll pay you sixty dollars a day."

"Thank you," I said. "I'll be happy to talk to your man from Munich, but please understand, I'm not doing this for money. I don't want to be paid for something that I feel is my moral responsibility."

"Don't worry about it," he smiled. "We all do our jobs and we're all paid for it. There's nothing wrong with money, and there's no conflict with your moral values. We're not giving you this money for your information; we're merely paying you for your time." So at the end of the day, he gave me sixty dollars in cash. It was the first money I made after leaving the Soviet Union.

The next day, at 8 o'clock in the morning, I went again to the embassy and was introduced to a tall blond man about forty years old. He spoke very good Russian, though with a strange accent. It was an accent I knew well, for it was how inhabitants of the Soviet republic of Georgia spoke Russian. I asked him, "John, how come your Russian has such a funny accent? You're not Georgian, are you?"

John laughed. "Actually, I'm an ethnic Irish, but I studied the language at Columbia and my professor was Georgian. So I picked up his accent."

John not only worked for the intelligence branch of the U.S. government, he also was a very intelligent man himself. I enjoyed talking with him. He had an excellent knowledge of life in my former homeland, even though he had never visited the Soviet Union. At one point, wishing to impress me with his knowledge, or perhaps hoping to catch me off guard, he asked, "By the way, how is Nikolai Alexandrovich feeling now?"

I did not understand to whom he was referring.

"Don't you know Nikolai Alexandrovich Semikhatov, the principal designer of P.O. Box 320? Just a couple of months ago he had a serious operation."

Naturally, I had heard the name from my former schoolmates who worked at P.O. Box 320. Ten years earlier I myself had narrowly escaped being hired by that same institution. Semikhatov was already head of it at the time. Now, in 1993, as I am writing this story, I have seen him on CNN in news coverage from the former city of Sverdlovsk, which has now reassumed its original name of Yekaterinburg. Semikhatov still runs the same facility, though his face has now openly appeared on American television. Today he is a public figure trying to raise money for the Russian economy. But two decades ago his very name was a secret, let alone his position, residence, or the projects he managed. I remarked to John that his sources were better than mine, for I had heard nothing of Semikhatov's illness. In fact, I had never met the man—I was too small a potato to get anywhere near him.

Next we talked about my adventures at the P.O. Box. John showed me a large satellite map covering the area where the secret facility was located, and asked me to relate the images on the map to my personal impressions. I do not have a photographic memory, but the emotional intensity of that remarkable summer day so amplified my senses that when I closed my eyes I could see everything as if it were a movie.

God works in mysterious ways and the twists of fate can sometimes be quite amazing. Just a short time before, I had been arrested on the premise that I was an American spy, though I was not. But now, here I was sitting at the U.S. embassy in Austria with a high-ranking CIA officer, sharing with him the results of my observations in that secret facility. In fact, Colonel Vereshchagin had been right. I *was* an American spy, even if I did not yet know it myself.

John was very pleased with my recollections and we worked together for several days. Then he returned to Munich, and I had no further contacts with the CIA for several years.

My next encounter with the CIA was in September of 1980. I received a call from a CIA agent in Boston whose real name was Ralph Kasperovich and who asked me for a meeting in New Haven, Connecticut, where I was lived at the time. I told him to come on over, that I was glad to be of help.

We had just bought our first house and our daughter Julia was only one month old. I was painting the ceilings and hanging wallpaper. Five-year-old Roman was constantly jumping and twirling about (his figure skating talent was already in evidence). It was a very busy time, indeed. When the agent arrived at the house, it was difficult to find a clean and quiet place to talk.

Ralph brought with him the same map of Sverdlovsk that John had shown me in Vienna three years earlier. To my surprise, it did not include any of the corrections I made then. Probably it was just an old print. Ralph asked me to clarify one interesting question.

In 1979 a leak from a bacteriological warfare plant in my former hometown released a deadly anthrax virus. Reportedly, many civilian people were infected and a fair number died. It was obvious that the infection had been artificially produced as a bacteriological weapon, and the CIA was interested in pinpointing the site of the leak and the possible location of the production facilities. I told Ralph that the outbreak of the infection had not occurred until some two years after I had left the country, and that I could not possibly know anything about it. He said he simply wanted to know whether I had any idea where it could have happened, or perhaps knew anything at all about such places where bacteriological research was conducted?

I then recalled that about five years earlier some former colleagues and I had been cross-country skiing in the southwestern suburbs when we came across a tall red brick building out in the woods surrounded by a barbed wire fence with guard towers at each of the four corners. One of the members of our group told me that this was the "flea institute," as the place was nicknamed. He explained that his girlfriend worked there as a nurse and that he learned from her about deadly infections being developed behind those walls for military purposes.

That was all I knew about the place, and with some difficulty I was able to find the spot where the brick building stood on the map and showed it to Ralph. But when I did so, the agent squinted his eyes and said, "Why are you lying to me? We know from reliable sources that the leak occurred somewhere in the southeastern area of the city, not the southwest as you would have me believe."

At his words I became very angry. "Look here Mr. Kasperovich, I told you that I knew nothing about the leak. I was living here in the U.S.A. when it happened. I am just telling you where a *similar* facility was possibly located five years ago, just as you asked me."

He insisted, however, that I was deliberately hiding something from him and was avoiding telling him the truth. Apparently, he had expected more information from me and had no place else to turn or anyone to ask. As far as I know, at that time I was the only recent immigrant from Sverdlovsk living in America. The agent left my home quite disappointed.

A few days later, I received from New York a copy of the American Russian Daily, to which I then subscribed. There on the first page was a map of the southeastern area of Sverdlovsk and a detailed description of the anthrax accident: how and when it had occurred, how many people had been infected (a total of 64 died), and so on. I took some scissors, clipped out the article, sealed it in an envelope and sent it to the CIA office in Boston. Two days later Mr. Kasperovich called me in great excitement. "That was fantastic information! The CIA didn't know these interesting details. It's very valuable. Thank you so much!"

"You are welcome."

I did not know whether to laugh or cry. That "valuable" information had been known to anyone who read a Russian newspaper published in New York, but was unknown to the CIA. So much for efficiency.

Several months later Ralph called again and asked me if any of my relatives had mentioned anything to me in letters or phone conversations or given any indication of what was going on in Sverdlovsk, or anything about the medical nature of the accident. Suddenly, a terrific idea came to me.

"Ralph," I said, "I understand that you need that information right away, but what would you say if I offered you a way to get more *inside* information—though with considerable delay, maybe a year, maybe more?"

"What are you talking about? Is this for real?"

"This is not a conversation for the telephone. When can you come down here?"

The agent said he would catch an Amtrak train the next morning and would be in New Haven around lunchtime.

The following day, I took some time off from work and met Ralph at my home. Irena watched Julia and Roman so that I could talk to Mr. Kasperovich without interruption.

"Look," I said, "I'm not being entirely unselfish. We can both get something here with very little effort and a bit of creativity. My wife's brother Misha Tyshkov is an M.D. in Sverdlovsk. He works in an hospital emergency ward. I think that he may know a great deal of the sort of information you are looking for, especially of a medical nature. Would you like to talk to him?"

"Of course we would. But it would be too risky, both for our agent in Russia and for your brother-in-law."

"Ralph, I'm not talking about meeting Misha *in* Russia. First, we must try to get him *out*. I mean legally, without raising any suspicion."

"Does he want to leave Russia?"

"Yes, I am positive about it."

"Keep talking," said the agent. "I'm all ears."

To make everything clear, I first had to explain my own story. Ralph had no doubt read my file, since the circumstances of my escape from the P.O. Box were generally familiar to him, but he did not know the full details.

When I came to the end of my introductory speech I said, "If Colonel Vereshchagin is not yet retired and his whereabouts can be discovered, we might try to blackmail him. I could write him a letter asking him to arrange for Misha to leave Russia, much as he did for me. In exchange, I will promise not to go public with the story of my visit to the P.O. Box and his assistance in my emigration. I don't know if it will work and how long it might take, but what do we have to lose? The only thing your people in Moscow would have to do is to drop my letter into a regular mail box somewhere in Russia, so there should be no real danger for them. With respect to Misha, he will know nothing, and if everything fails, he will just be an irrelevant pawn in the game. What do you say?"

The agent scratched his head. "For now, I'll say nothing. It's not my department, but I will pass your suggestion on to the appropriate party." That was all I wanted to hear from him.

A few weeks later, Ralph called to say that he would like to visit me again with a man who wanted to talk to me about the plan. When they arrived, this man gave me some good news. He said that my old "buddy" Vereshchagin was still posted at the same P.O. Box and that he still held the rank of colonel. The man said that my plan was interesting, though several people at the CIA had looked at it and thought it had little chance for success. Nevertheless, given that the Soviet Union had invaded Afghanistan and the U.S. government was very much concerned about the possibility of the Russians using bacteriological weapons, they had decided to give it a try. The CIA was willing to exploit any possibility to gather more information and hoped that through my plan they might do so with minimal effort.

At this point, it was up to me to write a short letter to the colonel. My visitor opened his leather briefcase and took out a few Soviet envelopes with postage stamps, a Russian ballpoint pen and a couple of sheets of similarly authentic paper.

I wrote a couple of drafts, which my visitor approved after a number of corrections, and then I copied out a final version. It was short and in English translation went something like this:

> "Dear Comrade Vereshchagin,
>
> "I trust you still remember our last brief chat at the bakery store. I am so grateful that you brought me my camera, which I thought was lost forever. We are now living quite well, the kids are fine, and I hope that everything is okay with you, too.
>
> "My wife misses her brother very much. So I thought maybe you could do for him and his family the same thing you did for me, so that they might be able to move closer to us. As long as he's not here I find it increasingly difficult to keep my word to you. Let us be of help to each other. You help him and I will be as cooperative as before. Write and tell me your decision. My address is: Moscow, Post Office number so-and-so, General Delivery."

I signed the letter with my real name.

I wrote the P.O. Box address and the colonel's name on the envelope and put the Moscow return address at the bottom. We made the letter fairly direct because we had little worry that it would be opened by a censor. The letter looked like genuine domestic mail and should only be checked by a P.O. Box censor reporting to the colonel himself. We hoped that the censor would not dare to open a personal letter addressed to his boss, and indeed, that was the biggest risk we had to take. My visitor said that the letter would be dropped into a Moscow mailbox in a few days, and that all we had to do now was to wait and see whether the colonel responded.

After a month or so the CIA man called from Washington and said that he had some encouraging news for me, which would be delivered by Ralph. Needless to say, I was delighted when Ralph brought me just five words typed on a sheet of paper: "Agreed, but not from Sverdlovsk." There was no signature. We understood that it would look too suspicious for the colonel to arrange an exit for Misha from a city from which emigration was virtually nonexistent at the time. No doubt he wanted to buy some time, or just wished to play it safer.

Communications with most Soviet cities were very difficult during those years of the Afghan war, and it was only after considerable effort that I managed to get a call through to Misha and tell him that they had to give up their apartment for a residence in some other city. Naturally, I did not tell him the reason, but he knew in any case that his chances were better someplace else. After some time they moved to the town of Mga near Leningrad. The name of that town can be translated as "gloom," and it certainly lived up to its name. But no matter—with luck, Misha, his wife Masha, their newborn son Alex, and Irena's parents would not have to be there too long.

Now my task was to arrange a fake invitation for them from Israel. The CIA could be of no help in this. Instead, I went to an underground dealer in New York. For a thousand dollars he sold me an authentic official invitation for Misha and his family to come to Israel as permanent residents. The invitation was then smuggled to Leningrad (which the New York dealer also arranged, for additional fee) where Misha picked it up. They then filed their petition with the local authorities without delay.

My next and last joint effort with the CIA in this enterprise was to write another letter to the colonel to let him know the new address of my relatives in Mga. It was clear that the colonel was very well connected, for within a short while Misha's family received their permission to leave Russia for good. This was indeed good fortune, for at that time the Soviets were limiting emigration to just a few hundred people a year. When Misha arrived to Rome, he was interviewed by the CIA agent there. I do not know if his insights were deep enough and met the CIA's expectations, but I am positive that whatever he knew was worth what little effort the U.S. government spent to get him out.

Today Misha lives with his family in New Jersey and works as a director of pediatric gastroenterology at the Overlook hospital. As for the colonel, I have no idea what happened to him. Most likely, he decided to retire and thus spare himself any further consequences of receiving another letter from an old "buddy."

I had another encounter with the CIA about a year later. The Soviet invasion of Afghanistan was escalating when I received a call from Washington. A man by the name of Steven asked me for a meeting. I met him and another man, David, in a motel near New Haven where they were staying. Steven was an analyst, while David, who spoke perfect Russian (his parents were Russians), was something between his bodyguard and a bouncer. As Steven and I sat in chairs in Steven's room, David lay on the bed watching Bugs Bunny cartoons on TV. To amuse himself and, I presume, to make me more cooperative, he played with his gun.

Steven wanted me to give him my assessment of the efficiency of Soviet military technical personnel under combat conditions. I explained to him that my only personal experience with anything close to a combat situation had been a long time ago, during the Six Day War of 1967 between Israel and its Arab enemies. At that time I was stationed at the Soviet missile division near Sverdlovsk for my military field training. The entire war machine of the Soviet Union was placed on high alert. The Ministry of Defense wanted to take advantage of this moment of extreme tension to test the readiness of its air defense arm. As an exercise, low-flying planes were sent toward our radar station and the division commander was ordered to fire simulated missiles. These "missiles" did not exist in reality but were just electronic flashes on the radar screens. The commander was ordered to act as if this were the battle field. To verify that his division had performed well and "shot down" all the targets, the master radar monitor had to be photographed during the exercise, and the film together with a report were to be sent to the regional headquarters. I was assigned to take these photographs.

I was sitting in the control vehicle next to the commander. The inside of the car was dark. The only light came from green monitors that showed images of the sky. These images were produced by signals from a powerful rotating radar antenna located on top of a nearby truck. The commander was extremely nervous, and was constantly shouting, swearing, and cursing. When the planes began flying over our heads, he and his people panicked and could not do anything right. In violation of all the rules, he sent his assistant outside with a pair of binoculars to count the planes flying above. Though the radar station was equipped to detect and fire the missiles toward low-flying targets, the monitors showed nothing meaningful to be photographed for the report. If it had been an actual combat situation, we would all have been dead by then.

When it was all over, the commander was in a terrible state and did not know what to do. By this time, I had already spent three weeks in his division and was sick and tired of it. So I told him that if he could help me, I would help him. I promised to film an entire air combat situation simulating his great victory. In return, I asked him to let me go home before the end of my long field training. He agreed readily, and I proceeded with my part. I had a pretty good knowledge of how to create special effects for film, and it was no big deal for me to simulate plane flashes on the screen and superimpose on them the blips from the "missiles" and other appropriate indicators on the monitor, such as the time, coordinates, and so on. The commander was very impressed when I gave him the finished film. He kept his promise and I went home, coming away with scant respect for the efficiency of that particular air defense division in a combat situation. Ironically, this was the same division that had shot down Francis Gary Powers seven years earlier.

I told all this to Steven. I also offered him a good deal of additional information which I knew at second or third hand. Everything supported my opinion that the Soviet military machine was not terribly efficient when faced with a well organized adversary[1]. When Steven heard this, he became very angry. He shouted that I was intentionally trying to diminish the effectiveness of Soviet power in order to mislead the CIA. Apparently, he expected me to say that the Soviet war machine was superior to the American, and that its actions could be devastating for any potential adversary. It was as if he already had all the answers and only needed me to provide him with supporting facts. The Carter administration was trying to force the Congress to spend more money on the military and had ordered the CIA to come up with evidence that the American army was weaker than the Soviet. I became increasingly irritated by Steven's obvious lack of interest in the truth. He was merely fishing for anything that fit his preconceived picture. The two of us parted none too pleased with one other.

---

[1] A few years later my assessment was confirmed by the virtual defeat of the Soviets by the poorly organized Afghan rebels.

My last meeting with the CIA was far more affable. It was 1985. Gorbachev had just come to power and the Reagan administration wanted to develop a strategy for dealing with the new Soviet leadership and its mystical "perestroika." The CIA was directed to prepare its recommendations. Somebody somewhere in that organization came up with the interesting idea of organizing a conference on the prognosis of the development of the Soviet state in those turbulent times. It was decided to invite just five people with special backgrounds to the conference. These people were all to be entrepreneurs with extensive insight into the Soviet politics, industry and science, as well as a good understanding of the economy of the West. The expectation was that these five people would be able to come up with some recommendation that would give the U.S. government a useful method for dealing with Gorbachev and the new wave in the Soviet politics. At least, that was my understanding of the purpose of the conference. I was one of the five people invited.

On April 13, I flew in a tiny plane from New Haven to Washington's National Airport. There I was met by a CIA man who spoke almost perfect Russian. We took a van to a Washington suburb and stopped in front of a brown two-story office building. The sign on the door read "The Universal Insurance Company." It turned out to be just a front. In a large room I was introduced to four other people, all Soviet émigrés. They were probably the most interesting people from my former homeland I had ever met in this country. They were all inventors and entrepreneurs, quite successful in their lives and their American business ventures.

The room contained a large round table and a closed-circuit TV camera on the ceiling. We were introduced to a group of analysts and several agents who all spoke Russian very well. Though everyone apparently knew English, the entire conference was to be conducted in Russian. A man who introduced himself as a deputy director of the CIA told us not to pay any attention to the TV camera since it was broken (right!). We were given a list of topics to discuss, and for the next six hours or so, with a lunch break at a nearby restaurant, we conversed on these subjects.

The CIA people sat on couches and chairs in the corners of the room, continually taking notes. Now and then throughout the day, long lists of new questions were brought in. We were each asked to answer these questions individually, then our answers were immediately fed back into a CIA computer in Langly. The computer crunched our responses and compared them with existing U.S. government positions and policies on the specific issues. Then the results of the comparison were brought back to us and we had to clarify and extensively discuss whatever discrepancies had been found. Sometimes, we were asked questions by the analysts, who listened carefully to our deliberations. In contrast to my unpleasant experience with the CIA a few years before, these people were searching for the truth, whatever that might be. They never tried to steer us in any particular direction. If our opinion contradicted their previous notions, they simply asked more questions and took more notes.

After the meeting, I asked the deputy director about this surprisingly pleasant change from what I had noticed in the CIA agents five years earlier. He smiled and said that the CIA had passed through a difficult period and that these were new people. Even those who had worked for the CIA for many years, were of an entirely different breed. In the old days, they had remained in the background, out of sight, as the previous administration had little use for them. I hoped that he was right.

Unlike many Americans, I felt a great respect for the difficult, often dirty, but unfortunately essential job that the CIA was doing. We live in a turbulent and vicious world. It truly is a jungle out there—not at all a place to wear white gloves. That is the sad reality, whether we like it or not. Just as we respect our armed forces who fight openly, we must also esteem those who fight in secret, for they all fight for the same noble cause.

7
Starting Over

> Happiness is essentially a state of going somewhere, wholeheartedly, one-directionally, without regret or reservation.
> -William H. Sheldon

On leaving Russia, we cut all our ties with the past, our friends, our habits, with museums, theaters, our favorite outdoor spots and places to walk. Emigration from the Soviet Union was a one-way journey with no return. No one expected any of us to come back in our lifetime. For those who remained behind, it was as though we had died, though we were still very much alive to ourselves. Like ghosts, we slipped through the iron curtain without lifting or breaking it.

In 1976, just for the fun of it, I made some economic calculations of the Soviet system. It turned out that the Soviet economy would be able to support the lifestyle of the Communist party apparatchiks for about ten more years. As soon as these "servants of the people" could no longer live in their posh apartments and reap lavish profits from their positions, I thought, the system had to collapse. As it happened, I was not far off in my predictions. Gorbachev came to power in 1985, and the Soviet Union collapsed in 1991.

But in 1977 I had little faith in my own prognosis and left the country of my youth with the sense that I was moving to another planet. A farewell party with my friends and relatives seemed more like a wake. I had the feeling that even if seemed more like a wake. I had the feeling that even if someday, far in the future, I was able to come back, the cosmic clock, which runs at a different pace, would mess everything up. If I returned, I would find no one alive in the old place. It would be a totally different and unfamiliar world. For many years of my American life I did not even consider revisiting my former homeland, and to this day I have not returned. If I do go back, what will I find there? Most of my friends are gone. Some have died before their time, others have emigrated and are scattered throughout the planet, while others still have forgotten all about me, and I about them. My parents and my sister live in Connecticut. What would I see in Russia, were I to go? There is a different life there now. New people, new feelings, new interests—even the language has changed significantly. I do not know them and I cannot miss what I never knew. Nobody waits there for me, nobody needs me. I do not believe in nostalgia for places. I do believe in nostalgia for the past and one's youth. But it is a useless sentiment, for time, like the exodus from Russia, flows in one direction only, and there is no point in looking back nostalgically. The world has changed, and my home is in America, a country I love dearly and admire greatly.

Our new life in the West began the moment we landed at Vienna's Schwechat airport. We spent four enjoyable months in that country before moving to America. In Austria, we found several good friends who made our stay there one of the most pleasant times of our lives. As we had no work permits, the only money I was able to make came from my short stint consulting for the CIA.

Oh, yes, I also made about $120 by helping to improve state of health of one very rich woman. That funny episode deserves a brief description here.

In the seventies, all transit émigrés traveling to Israel were separated from the other new arrivals and placed under heavy guard by the Austrian police to prevent Arab terrorist attacks. They were then quickly flown from Vienna to Tel Aviv. Of the other émigrés, those bound for America, Australia, or Canada were transferred to Rome after several days to be processed by a consulate of the appropriate country. Those planning to stay in Europe or going somewhere else remained in Vienna and lived in overcrowded and filthy hotels. Thus, all such vagabonds had their own Viennese or Roman holidays. In rare cases, these holidays lasted just a few days, but more often they continued many months.

Vienna's transit hotels were run-down establishments owned by a brilliant crook named Madame Bettina. In exchange for providing temporary lodgings to thousands of refugees, she was more than amply reimbursed by the international charity organizations. As a business, her hotel empire enjoyed an extremely high profit margin. Besides the millions she made from her hotels, she also would buy at cheap prices whatever she could squeeze out the poor souls who passed through her lodgings. She then sold these items through a network of stores she also owned. Many of those emigrating from the Soviet Union brought with them various small valuables they hoped to sell upon their arrival in Austria or Italy and make a few bucks. These items included ethnic souvenirs, small cans of caviar, cheap cameras, lenses, and so on. Madame Bettina bought everything. If a person had nothing to sell, she searched out whatever else of value he or she might have to offer, such as a particular skill or talent. If a man was a carpenter, she asked him to do some work on her buildings, paying him little more than a token wage. If he was a painter, she brought him paints and a canvas and asked him to paint her a picture. She cared nothing for art and did not understand painting. It could be anything: realistic, abstract, surrealistic pictures—it made no difference. She knew that she could always find a buyer and pocket the profit. As a rule, she would not have to pay the artist, for he already felt enriched by being able to work on a real canvas with real paints.

Somehow, Madame Bettina heard that I had studied Chinese folk medicine and could treat some stubborn illnesses with electrical stimulation of the body's acupuncture points. In fact, I really had made some studies of acupuncture and designed an electrical stimulator which sent electric currents through the skin without needles. It consisted of a pen-like probe attached to a control box. Of course, one had to be familiar with the maps of the acupuncture points on the human body and their effects on a person's state of health.

The device was as effective as traditional acupuncture, but with far less risk involved. Also, it was far easier to use and did not require the years of training that one needed to practice acupuncture with actual needles. I had managed to bring the device with me to Vienna. Since few of the refugees could afford a visit to the doctor, my little electronic gizmo became a popular home remedy. I used it to treat everyone I believed I could help. Naturally, I never charge anything for my services. I was not a doctor, and besides, what could one get from people who had nothing? (Madame Bettina, of course, would definitely disagree with that statement.)

By the end of August we had already moved from Madame Bettina's hotel and were living in a beautiful apartment in the center of the Austrian capital. One day I ran into Madame Bettina on the street. She was very amiable.

"Jake, my friend," she said. "I am so happy to see you! Look, I need your help. You must treat me."

"What's wrong with you, Madame Bettina?" I asked, trying to conceal my irony, for I knew very well that this was her standard pitch—to get something for free, even if she didn't need it.

"Oh, you must treat me for gluttony. You see, I have an enormous appetite, and whenever I see a fridge, I open the door, then open my mouth, and eat, eat, eat ... Can you help me?"

I suspected that her problem was a huge appetite for cash, not for food—but either way it would likely be responsive to a tranquilizing electro-puncture treatment. I said that I would try and she made an appointment with me for an initial session. After that first visit, there were twenty more. Each time she came to my apartment I spent nearly an hour sending electric currents through her legs, hands, ears and face. On the fifth visit she brought along her sixteen-year-old daughter who was on the verge of a nervous breakdown, and I treated the two of them. After about three weeks, Madame Bettina announced that she was satisfied, and that both she and her daughter had begun to sleep more and feast less. I too was pleased that my treatment had been so successful, and handed to her a typewritten bill for fifteen hundred Austrian shillings, the equivalent of about $120.

When she saw the bill, Madame Bettina was thunderstruck. "Are you crazy!?" she screamed. "I thought you didn't take money for your treatments!"

"I don't take it from those who have none. But I guess you can afford fifteen hundred shillings. Besides, it's not much for twenty sessions with two patients."

"It *is* too much! Give me a discount!"

"No." I decided to be adamant with the old skinflint. "If you can't afford it, I will take nothing. But if you can, there is no discount."

With trembling fingers she opened her handbag, counted out three five-hundred bank notes, and left very sobered.

The next day I was the talk of the town. The entire émigré community of Vienna was stunned by the news that, for the first time in living memory, Madame Bettina had paid money to someone else, rather than putting it in her own pocket. As for me, this encounter with her was my first lesson in rich misers who have no qualms about squeezing every penny they can from those who have next to nothing to begin with. In America I have met several such specimens, and on more than one occasion, unfortunately, have had no choice but to do business with them. It is a pity that we can so rarely afford to be choosy.

While living in Vienna, I continued studying French and learning about the Western way of life. Many things we had to discover for ourselves. The very first photograph I sent to my parents from Austria was a picture of Irena and me holding a coconut. Neither we nor our parents had ever seen an actual coconut before—it was extremely exotic produce to us and something to brag about. Likewise we had never tried pizza before, though I heard from people in Vienna that it was delicious. So one day I went to a local supermarket called Löwa and there in the frozen-food section found a box marked "pizza." As we had never had frozen food before either, I had no idea what to do with the rock-hard pie. After thawing it to room temperature and eating it cold, we decided that pizza was a hoax, or at the very least, overrated. The next day someone told us that once the pizza was defrosted, we had to bake it before eating it. So we went and bought another one, and there in our tiny room in Madame Bettina's hotel heated it on a frying pan. It tasted a little bit better, but still not great. Not until we got to America did I discover truly good pizza. When I visited Italy, however, which I believed to be the birthplace of pizza,[1] I found that pizza there was not as good as in America. In fact, I think that ethnic food in America tends to be far better than in many native countries, with the possible exceptions of France and China. Some may disagree with me, but that has been my personal impression. The best Japanese, Mexican and Thai food I have ever had has been in California, the best Russian food was in New York, and the best Polish food in Chicago. To be fair, however, I must mention I sampled perhaps the finest quality and variety of foods from all around the world in Singapore.

One day our friend Tanya, the wife of the internationally renowned cellist Boris Pergamenschikow and a refugee herself, asked me why I was waiting around in Vienna. I told her that we hoped to receive French visas and move to Paris. She gave me a strange look.

"Are you out of your mind? You're not a painter, or a sculptor, or anything like that. What are you going to do in France? That country is no good for you. You must go to America, because you belong there."

"But I know nobody there. And I know nothing about America either. I have read many books about France, though. I know its cinematography, its art, its culture..."

"That's the reason I say you must go to America! You are not an artist or a film maker. You are an engineer and America is the right place for you. Find someone there and ask him to find you a job."

I thought a lot about what she said and finally decided to give it a shot. I sent my resume—or actually a scientific version, a *curriculum vitae*—to several American universities. While living in Russia, I had exchanged reprints of technical and scientific papers with some American scientists. I had with

---

[1] The real birthplace of pizza, as I was told, is Greece.

me only five names with addresses, and therefore sent out only five copies of my *vitae* to those people. I also included reprints of my papers, with summaries in English, so that my correspondents could get some idea of my background.

Meanwhile, I stopped studying French and we began going to English classes taught by a former American pastor, Tibby Lorenz. Besides being an excellent teacher, Rev. Lorenz was a colorful person and a very kind and classy fellow. His lessons taught us not only the English language, but also about the United States in general and the American way of life. He was a very compassionate and generous man as well, and he used whatever he could spare from the little money he made teaching refugees to help out as many people as he could. He jokingly called himself Dracula, remembering probably his Transylvania roots. Tibby was also a dandy and loved to wear stylish and striking outfits. Every day he wore a different suit, with bright shirts and neck kerchiefs. His hair was wavy and shiny, and he had a long and old-fashioned mustache, despite his youth—he was barely over thirty. And indeed, he looked very much like Dracula. I thought to myself, "Wow, what a strange country America must be, if a man of the cloth wears such clothes!"

One day he came to our apartment and said to me, "I don't like the way you dress. Show me your suits."

When I opened the armoire, he was stunned. "What! That's all? You have just ONE suit?"

With that, he dashed off, and half an hour later brought me back three jackets and two pairs of slacks from his own wardrobe. It was no use refusing the gift or trying to tell him that none of the clothes were my size and there was no way they would fit me. Tibby did not want to hear it—he just wanted to make me the present.

Thanks to Tibby, my English was improving noticeably. Before we left the Soviet Union, I could read and write some English. In school I had studied German and French, though with little conversational practice, of course. Getting hold of any foreign books, magazines, or newspapers was virtually out of the question, except for those published by Communist organizations. But who wanted to read those dull tracts?

Owning non-communist publications, though not a crime, was still rather dangerous.

I had a friend in Moscow who often mingled with the foreign press folks. Now and then, they supplied him with German, French, and American magazines. Somehow, he managed to get away with possessing such risky publications. He tried, however, not to keep too many of them at his home, and one day he gave me two American magazines, the most important publications, he explained, of the free world. One of them was *Playboy* and the other *Penthouse*. The pictures in the magazines, especially the centerfolds, were very impressive, and I thought, "These Americans are really good-looking people!"

Some of the pictures were very puzzling to me. One centerfold showed a beautiful blonde. A few parts of her body were symbolically covered by some kind of sports outfit. In one hand she held a strange-looking elongated item whose purpose was a mystery to me. It was something between a hockey stick and a billiard cue. One evening, I gathered a group of my closest and most trusted friends, pinned the centerfold to the wall and offered a prize to whoever could correctly guess the intent of that strange stick. It was obvious to us that it was something very kinky, but for what purpose?

After an hour of unsuccessful attempts, the prize was still unclaimed. I decided to read the text in the hope of figuring it out. For any study, one needs an incentive, and this puzzle was a great incentive for me. I got a dictionary and some textbooks and starting studying. That was how I began learning English. Not surprisingly, my initial vocabulary acquired from those two magazines, was, so to speak, of a fairly particular nature.

Ten years earlier I learned Esperanto of curiosity and enjoyed its grace and logic structure. English, however, was million times more difficult and enigmatic. Nobody could explain to me why *sharp* and *blunt speech* are the same, why *overlook* and *oversee* are opposites, why a nose can run and feet can smell, if a *vegetarian* eats vegetables, what should eat a *humanitarian*? Spelling also was a struggle. Especially, I had problems with words which came from other languages. Why, for instance, you need *p* in word "psychic?" *It pcertainly does*

Starting Over 151

pseem psomewhat psilly. Nevertheless, I enjoyed studying English—after all, it is the most versatile language—about two million words! Does anyone know at least a half of that number?

When we applied for emigration, Irena and I began taking more serious lessons in both French and English from a retired school teacher, but I received my real taste of American English from Tibby.

Oh, yes, the girl in the centerfold was holding something I had never seen before—a golf club.

A month or so after sending out my resumes, I received three letters. One was from Hartford of Connecticut, another from Palo Alto of California and the third from San Antonio of Texas. The three university professors regretfully informed me that they could do nothing to find me a job. Another two weeks or so after this, I found two more letters in our mail box. One was from Paris. It contained our long-awaited three-month French guest visas. The other envelope was from Cleveland, Ohio, where Dr. Michael Neuman offered me the position of a postdoctoral research fellow at Case Western Reserve University at a salary of $12,000 a year.

I was stunned. I could scarcely believe my luck. First, I ran to Tolstoy Foundation and told Mr. Rogoiski that I want to change my destination to America, instead of France. He told me that it would be no problems as long as I had a job offer. But getting a U.S. visa would still take at least four to six months.

The next person I told about the letter from Cleveland was Tibby. He said, "That's fantastic! There've been very few cases of refugees being able to find a job before even arriving in the country. Don't pass up this opportunity. Besides, twelve thousand a year! That's a lot of money!"

Tibby was very excited. "Look," he calculated. "Rent on a two-bedroom apartment will cost you no more than $75 a month. Then food, that's another $50. Add to this clothing, transportation, movies and so on, and you'll hardly spend more than $300 a month. With that, you'll even be able to put away some $700 every month. That's why I call it a lot of money!"

Getting ahead of myself, I can say that his estimates were about four times lower than the minimum cost of living in Cleveland in 1977. It turned out that the last time Rev. Lorenz had been to America was in 1969. The value of a dollar had definitely changed a bit since then.

I filed a petition with the American consulate and prepared myself for a long wait. Meanwhile, many of our friends had already left for their resettlement destinations. The cellist Pergamenschikow, his wife Tanya, their son Danilka, and his mother, boarded a train and moved to Cologne, Germany. He is still there, working as a professor in a Hochschulle and giving concerts all around the world. Another friend of mine, a Lithuanian, left for America and promised to write me a letter with his initial impressions. Two weeks later, I received a postcard from him with a very short message: *"Folks, this is such a normal country!"* At the time, I did not understand what he meant by "normal," but when I saw America for myself, I agreed with his assessment. Compared to the surrealistic world from which we had escaped, America was strikingly normal and sane. Now, many years later, I can also see a great deal of insanity in America, but everything depends on one's point of reference.

To our tremendous surprise, less than three weeks after filing our application with the consulate, we received a telephone call from Mr. Rogoiski at the Tolstoy Foundation who told us that visa had been granted and we were scheduled to fly to America on December 19. No refugee had ever received permission that fast, and at the time I assumed that the processing of our application had been expedited thanks to the CIA, in gratitude for my input. As it turned out, it was not CIA who helped me. (Did CIA ever helped anyone?) It was Dr. Michael Neuman, my future boss, who had called his acquaintance Dr. Freeman (about whom I shall say more in the next chapter). Dr. Freeman had a friend named Milton Wolf, a Cleveland businessman who after Jimmy Carter's election was appointed the American ambassador to Austria in exchange for favors he had done for the Democratic Party. Dr. Freeman called his ambassador friend, and the latter set some wheels

Starting Over

and gears in motion and our application was swiftly processed.

Once we learned our departure date, I sent a letter to a lady in New Jersey, a cousin of my father-in-law, though we had never met. Her name was Elaine. Like both my wife and my father-in-law, she was a violinist. I wonder if that profession is hereditary? At one time Elaine had been married to the pianist Teisher from the Ferranti & Teisher duo, and she had retained her double last name of Sutin-Teisher. She was a first generation American. Her mother Henrietta, my father-in-law's aunt, had come to America, passing through Ellis Island, in 1910. Hoping to stay in New York for a couple of days on our way to Cleveland, we asked if Elaine could meet us at the airport.

We turned another page of our life with joy and great expectations. On December 19, 1977, we boarded a TWA jumbo jet bound for New York. When we landed at Kennedy ten hours later and passed through immigration control, we were met not by Elaine, as we had hoped, but by Mr. O'Connor, a representative of the New York branch of the Tolstoy Foundation. He gave us our airline tickets to Cleveland, $60 in cash, and said, "Good luck, guys. From now on you're on your own. Don't call us and we won't bother you either. Just don't forget to return the money we spent on your move to America."

With that, he left, and I walked out of the terminal to find a taxi. To our great surprise, the only language the driver could speak was Russian. What a country! We told him that we needed to go to Englewood, New Jersey. He said that it normally would cost us $30, but since it was out of state, the fare was twice that, or $60. Later, we were told that he had lied and illegally charged us double. A fine compatriot he turned out to be! The money we paid him was about a fifth of our total resources, but we had no choice.

When we arrived at Elaine's house, we discovered that our letter from Vienna had never arrived and she was not expecting us. But Elaine went out of her way to make us comfortable and to pacify three-year-old Roman who was deathly tired from the long journey around the world and seemed on the verge of tears.

We spent three days with Elaine and her parents. She drove us around New York City, which was soaked with rain but nevertheless was aglow with the breathtaking Christmas lights. Being a musician, the first place she took us was Carnegie Hall. New York stunned and enchanted us all at the same time. On December 22, Elaine brought us to Newark airport and we took a flight to Cleveland, a our new home.

At Cleveland's Hopkins Airport we were met by Dr. Mike Neuman, who had sent me the letter with the job offer. Today, so many years later, I still feel deeply indebted to that man whose letter changed our lives forever. Dr. Neuman and his secretary Lois Schweizer helped me rent an apartment in Cleveland Heights, not far from a bus stop, so that I would be able to commute to the university. Mike even brought to us a number of basic necessities: dishes, silverware, some clothes and even a small TV. This kind of a goodwill and compassion had long been forgotten in my old country, and we were deeply moved by such unexpected support.

Mike invited us to his house for a Christmas party. There I met his wife Judy and daughter Elizabeth, along with many other members of his family. Mike's father asked me if I would like some root beer. I said, sure, I love beer. When I tasted it I nearly choked. It was the lousiest beer I had ever tried in my life! Far worse than anything I used to drink in Russia. Ever since then, even after having discovered real beer here, I still cannot stand the smell of that soft drink. That's how great my disappointment was. But our warm and cordial welcome by all those people was very important for us at the beginning of our new life.

A few days later, we heard a knock at the door and opened it to find a strange-looking couple standing there. The man had a long beard and the woman's head was covered with an old-fashioned kerchief, which Americans call a "babushka" (the Russian word which actually means "old woman"). They told us that they were representatives of the Chabad, an ultra-religious Jewish sect of which I never before heard. They offered us some spiritual literature and invited us to visit their temple. The man said that he was a medical doctor and that I should call him whenever I needed help. He gave me his card with both his work and home numbers.

As it happened, just a few days after their visit, Roman became very sick. I suspected that his immune system had not had a chance yet to adjust to the viruses and germs of the New World. It had been an extremely severe winter, and Roman had caught cold while playing outdoors. He felt very ill and was having trouble breathing, which worried us greatly. I did not have the slightest clue what kind of medication to give him. In Russia, I would have known where to go and what to get, but we had been living in this country for all of a week and even very elementary things were still great puzzles to us. It was Saturday morning, the university was closed due of a blizzard, and there was no one around to turn to for advice. Then, suddenly, I remembered the bearded doctor who had visited us just a couple of days before and offered his help whenever we need it. I found the card and dialed his home number. No answer. I called again, and again. And again.

Finally, a woman answered. "Yes?"

"I want to speak to the doctor," I said.

"He can't come to the telephone. Today is Saturday," she replied, and was about to hang up.

"Hold on, please!" I cried. "If he can't come to the telephone, just ask him what medication I should give to a child who has a very severe fever, and where I can get that medication?"

"He can't tell you that because he doesn't work on Saturdays. If you need advice, go to a drugstore and ask there." And she hung up.

I walked to a nearby Revco drugstore and asked the girl at the counter what I should buy for my child, and she suggested children's Tylenol. It worked quite well. Before long, Roman felt better, and in a few days he had recovered completely. But never again did I call that bearded doctor or go to his temple. I do not even believe that he was a truly religious man, for no religion can forbid a man from helping another human being.

On January 2nd, I arrived at the campus to start my new professional life. There I was greeted by Professor Ko, the director of the university's Electronic Design Center, where my laboratory and office were to be. Besides Dr. Ko, there were a number of Chinese teachers and students at the university. I found them to be hard-working and tight-knit group of people, who were very supportive of one other. Being in this Chinese environment did not, however, do much to improve my English. Within a few months, someone told me that I spoke English with a Chinese accent.

I worked at the university with joy and great pleasure, arriving early each morning and leaving at night. It was strikingly similar to my routine in my former life: the same project and virtually the same equipment. But something was profoundly different, though at first I could not put my finger on it. Why did I feel so content, so relaxed, so peaceful? Then I realized that the main difference was the psychological atmosphere. In the Soviet Union, every plant, every factory, every research institution, every university, and every laboratory had its secret KGB informers who regularly reported on everyone. On the basis of these reports the government decided the fate of each individual. Everyone knew about this system and learned to behave accordingly. In effect, we all were actors. Some were better at it than others, but we all did it. Just as players on a stage speak words that are not their own, we learned not to say what we meant. I loved the theater and enjoyed acting on the stage, but I always felt terrible playing a part in real life.

In Cleveland I suddenly realized that I no longer had to play at something. I could say what I thought and think what I liked. What a pleasure it was to be mentally relaxed! That was why I felt so good about working there and enjoyed being myself. It was true freedom, which many people understand narrowly, like just being out of jail. Later, I learned that people in America, too, very often and for various reasons do not say what they think. However, from the very beginning I decided to be totally outspoken and always to say what was on my mind. Some people find it quite refreshing, because what they see is what they get. Many others disapprove of my outspoken nature and find it too rude, unconventional, or simply eccentric. Well, I really don't care what they think. I am what I am.

Soon we experienced a taste of poverty. When we left Russia, we were only allowed to exchange $120 worth of rubles. In Vienna I made a bit of money doing some consulting work for the U.S. government and treating Madame Bettina for gluttony. After paying our few Austrian bills, we arrived in Cleveland with about $300 remaining—$275 of which we immediately had to pay as a security deposit on our apartment. Since this was all the money we had, our kindhearted landlord agreed to wait a few weeks for the first month's rent. University employees were paid once a month, which meant that I could not expect my first paycheck until the end of January. But how were we to survive thirty days on just $25?

This money lasted us only a week, whereupon I took a bus to the Jewish Family Service which arranged for us an interest-free loan for $300 which we had to pay back within two months. This money kept us afloat till my first paycheck from the university. Naturally, we paid the loan back within the two months. In addition, over the next several years we managed to pay back to Tolstoy Foundation everything it had spent on our support in Vienna and relocation to America. While I am grateful to all the good people and organizations who helped us, it gives me a good feeling of pride to know that we started with practically nothing, paid off all our debts, and now pay fairly high taxes to Uncle Sam and the state of California. I consider these taxes the never-ending repayment of my debt to this country which has been so good to us and become our true motherland. But can any son ever repay his mother? Perhaps that is why taxes are so perpetual.

We managed our small income very carefully, always struggling to make ends meet. But these ends kept shifting, becoming a kind of moving target. Irena tried as hard as she could to help resolve the family's budget problems. Every day she practiced her violin in the hope of joining one of the local orchestras. She also sent her resume to the music college inquiring about any teaching positions.

One day, she received a telephone call from the head of the string department who invited her to come play for him, for he wanted to determine her professional level. She made an appointment with him and for the next three weeks practiced day and night. We put Roman in kindergarten, I was at the university, and she spent all her time working on her program, polishing and fine-tuning every last detail. On the appointed day, I drove her to the college and she played for the man. When she finished, he said that he was very impressed and wished her all the best.

"Wait," she said. "What about the job?"

"What job?" He looked surprised. "We don't have any openings here."

"But why on earth did you ask me to come here and play for you?"

"Oh," he said, smiling sweetly, "I just love listening music."

Her practicing eventually paid off. She auditioned for the Cleveland ballet and opera companies and became a member of their small orchestras. Of course, these jobs were not full-time positions and paid little, but nevertheless it was a very important contribution to our family budget.

Although our start was slow and not without its difficulties, I was happy and optimistic. In preparing to emigrate from the Soviet Union, I had not expected any easy solutions and was ready for whatever hardships I might find in my new life. I did not hope for any help from anybody. I did not expect to find a job in my field, or to live in a decent apartment. Let alone own a car! I would have been happy with any job at all: sweeping streets, washing dishes, driving a taxi, working as a longshoreman. Anything would do, just as long as I was allowed to live in a free country. What I discovered in America far exceeded all my expectations. I had found a good job, we had food on the table and clothes on our back, Roman was in kindergarten, total strangers had donated infinite small necessities to us. I felt I had got so much more here than I could ever have hope for, and I was very happy. Now, many years later, I am glad that I came to America with that kind of attitude. It made our adaptation so much easier. We were more patient and never shied away from hard work. Since that time I have met many immigrants who have arrived in this country with the belief that in America the streets are paved with gold. These people have all been deeply disappointed, but I have no pity for them.

The winter of 1978 was unusually severe. The East Coast and Midwest of the United States were hit by a devastating record-breaking snowstorm which killed some 100 people.[1] When we arrived in Cleveland, there was already a good deal of snow on the ground. It was a dramatic contrast with Vienna, but nothing unusual with respect to the Urals and Siberia. However, I found the climate around Lake Erie even worse than in Siberia, or, for that matter, in any other part of my former homeland. In Siberia and the Urals, the air contains very little moisture and even a hard biting frost is not all that bad as long as you have warm clothes and avoid inhaling the cold air. Thus, even temperatures which often fell below -40° were considered normal and presented few problems, except perhaps for some machinery. In Cleveland, however, the air was so damp that even a temperature of just a few degrees below freezing got right into your bones, and I did not like that.

The snowstorm which began on January 25 was really bad. A perpetual dusk had fallen and the wind blew at gusts of over 100 miles an hour. Ground and sky melted together into one gray howling mass. The entire city sank into a stupor—people stayed inside, businesses closed, and very few cars moved through the snow-conquered streets. The snow crews did not even bother cleaning the roads and freeways, for the blizzard would recapture its position within minutes.

When my alarm clock rang at 7 A.M., I looked out the window and saw nothing but darkness and glittering snowflakes bombarding the swinging streetlight. I had no radio on which to hear the advice to stay home—besides, in Russia snow was never a reason for not going to work. So I got up, and soon I was standing at the bus stop. I had to wait a long time, but in the end the bus arrived and took me to the university. This

---

[1] I am writing this in January of 1994, when, sixteen years almost to the day, the East Cost has been hit again with a snowstorm of similar intensity.

was unfortunate, for the university was closed. The guard was very surprised to see me banging on the locked door.

"What are you doing here in such weather? Everything's closed, all the buildings. Go home."

I went back to the bus stop. Stoically, I waited for about an hour—but no luck. The bus that had brought me to the university was the last one before all service was terminated. What was I to do? It was getting much colder and the blizzard was raging. I decided to walk home. It was about five miles. Very few cars passed me during my four-hour journey. The snow pelted my eyes like bullets. I was frozen through and lost every sensation but one—the will to keep walking. Everything has its end, and eventually I made it home. That trip cost me ten days of a high fever. I caught a severe cold and had to stay in bed. But what can one do? When I recovered, the first thing I bought was a radio.

Before Elaine took us to Newark airport, she told us that also living in Cleveland—where she herself was born and raised—was another relative of hers, my wife's third cousin. In fact, his last name was the same as Irena's maiden name, Tishkoff. Gary was a violinist with the Cleveland Orchestra. Truly, it cannot be mere coincidence that relatives born on opposite sides of the globe and had never heard of each other, all chose the same profession. It must be something in the genes.

A couple of weeks after arriving in Cleveland, we found Gary's number in the telephone book and dialed it. At first, he thought we were pulling his leg, but eventually I managed somehow to convince him that we were genuine and he came over to our apartment with his wife Harriet, daughter Mira, and son Eric. It was an amazing and joyful meeting. Gary was a tall and handsome man with a deep pleasant voice. He took Irena and me to a symphony concert at Severance Hall, while Mira babysat with Roman. During the two years we lived in Cleveland, Irena and Gary often played various musical duets at either our home or theirs, which drove Harriet mad. She was a very jealous woman and, considering Gary's good looks, not without reason. In America, I have learned several euphemisms. One of them is that instead of saying that a woman is ugly, one should diplomatically say that she has a good personality. Well, Harriet had a tremendous personality! I do not know the details, but a few years later she and Gary had separated. He still lives in Cleveland and plays with the symphony.

When we lived in Russia, my father told me about his uncle, the brother of my grandfather. The story goes that when World War I broke out, both brothers were drafted into the tsar's army. My grandfather went to the front, but his brother said that he did not want to be part of that stupid war and defend the empire of the tsar. He packed a few suitcases and he and his young wife made their way through war-torn Europe to Marseilles, where they boarded a boat bound for New York. They arrived in America in January of 1915. It was snowing in New York and the frosty wind was blistering, so he said, "I didn't leave cold Russia to find myself in the middle of the American snow. Let's go south." So he and his wife hopped on a train and went to Philadelphia. It was snowing, too. So they kept going south until they reached South Carolina. There it was raining.

"Well," he said, "this is better, but not good enough." They got back on the train and continued their journey until the train stopped in Jacksonville, Florida. There it was sunny, warm and very pleasant.

"This is it!" exclaimed the fussy immigrant. "I want to live here."

They got off the train and made their home in Jacksonville. His wife bore him four sons and a daughter. When World War II broke out, three brothers were drafted into army. This was not the tsar's Russia—they were good American boys and all went to the front. One was killed on D-Day in 1944, but the remaining two came back and still live in Jacksonville.

My late grandfather had a photograph of his brother's family taken sometime in the thirties. No doubt, he received it in a letter from America and saved the picture in spite of the danger of keeping such "compromising" evidence of relatives living abroad. I took the photograph with me when we emigrated.

One day I decided to search for these relatives. But how was I to find them? I did not even know their names, apart from their surname, which was the same as mine. So I went to the Cleveland public library, picked up a Jacksonville telephone book and to my great surprise found four Fradens. I selected the first in the alphabet, Abe, and dialed his number. Abe Fraden was very surprised when I explained that I was very probably his relative. He had no idea that he even had a cousin in Russia and asked me whether I had any proof. I told him about the photograph and offered to mail it to him. He agreed. I sent him the picture and three days later he called me back and said, "It's us!"

He invited us to come visit him in Florida. That May, we flew to Jacksonville and met Abe and two of his brothers, Joe and Bernie, and his sister Rosie, who was then visiting from Rhode Island. We had a very good time, and my only regret was that in 1914 my grandfather had not acted as wisely as his brother.

On the third day of our stay in Jacksonville, Abe asked if I would like to visit his mother, who had come here from Russia in 1915. His father had passed away some years before, but his mother, now ninety years old, was still alive and was living in a convalescent home. He warned me that she was a bit senile and that I should not pay too much attention to what she might say. I agreed and he drove me to the home. I waited in the lobby while he went to the his mother's room to fetch her.

A few minutes later I saw the old lady coming down the corridor. As I approached her, she looked up at my face and suddenly said in Russian, "I know you. You are the brother of my husband. What are you doing here?"

Abe was stunned. He had not even known that his mother could speak Russian. It seems I very much resembled both my grandfather and father at the same age. In her mind, times, places, and languages were all confused, and after sixty-four years she recognized in me her brother-in-law and even remembered her long-forgotten tongue. Indeed, the human brain is a very mysterious machine.

I had been told that no one can consider himself a real American until he gets a car and a credit card. After living in Cleveland four months we realized that our life was tremendously crippled because we had no car, and I agreed that, indeed, an American without a car was like a bird without wings. All of our travel was limited by bus routes and the good will of our friends, who drove us places when we were really desperate. So we decided to buy a car—a used one, of course, or, as people would say now, a "previously owned" car.

This strange usage of the English language, incidentally, reminds me of the use of Russian in Soviet propaganda, which employed a great number of elevated or ennobling euphemisms for ordinary-sounding words and phrases. George Orwell in his famous book *1984* called such language "newspeak." Either we failed to notice it during our first ten years in America, or modern "newspeak" is a phenomenon of the nineties. Today, instead of using straightforward words, Americans invent surrogates in the hope that they will sound more presentable. Now they say "African Americans" instead of "black people." (What about white immigrants from South Africa—aren't they also African Americans? And what is wrong with just being black?) They say "sexual preference" instead of "homosexuality." (I like blonde women more than redheads—isn't that also a sexual preference?) Insurance salesmen call themselves "risk managers" or "financial planners." Instead of saying "she is fat", people say "she has a full figure." (Does that mean that my wife has an empty figure?) And a "previously owned" car means that it was owned by someone other than the car manufacturer or dealer—in other words, it is just a used car.

Our first "previously owned" car was big. No, it was not just big, it was colossal. It was a Buick Electra, and it had an enormous appetite for gasoline. In 1978 this was a serious limitation, for gasoline was in a short supply, thanks to the Arab oil embargo.

Three-year-old Roman once remarked that "in the beginning, the car was very small. Then, it started eating gas and grew big." We took our first long trip in that car. It was the summer of 1978, and we went to see New York and Washington, D.C. Passing through New Jersey, we visited Elaine and her parents, Henrietta and Joe. We saw the musical *The King and I* on Broadway with Yul Brynner, and I took breathtaking pictures from the Empire State Building. As I stood with my camera above the magnificent spectacle of New York, I thought of how I would have been promptly arrested had I attempted to take aerial pictures of a soviet city. From New York, we drove safely all the way to Virginia, then on to Washington, where we visited every single museum we could find. We went on tours of the White House and FBI headquarters, and headed back to Cleveland happy and fully satisfied with our first American road trip.

While driving down the New Jersey freeway, we suddenly heard something drop off the car and drag along behind us with a terrible rattling noise. Naturally, at that same moment we saw a cop who signaled us to stop. When we pulled over, we realized that the Buick's electric horn had fallen off and was still attached by a long wire to the back of the car. The cop said he had never known such a thing to happen to a car and let us go—after I cut the wire, of course.

About ten miles later, we experienced a more dramatic drop-off. This time it was me who fell off—the entire driver's seat collapsed backward. Fortunately, I managed to slam on the brakes and we stopped safely on the shoulder. I examined the seat and found that all the brackets bolting the seat to the floor had rusted through and there was nothing holding the seat in place. This was a serious problem. We had about nine more hours of driving ahead of us and no money to get the seat fixed on the way. What could we do? We found a practical solution. Irena sat in the back seat, planting her feet firmly on the back side of my seat to hold it in place. I got behind the wheel and tried not to lean back as we continued our journey. I felt as though I was seating on a see-saw. That was one hell of a journey! It took us about twelve hours of extremely careful and painful driving. By the time we got home, I had a terrible backache and Irena had "legache".

I went to a hardware store, bought an electric hand-drill set and bolted the seat back to the frame of the car, adding a couple of new steel brackets. In doing so, I discovered that the entire floor beneath the carpeting was virtually non-existent—it had rusted completely through. I had to put a piece cardboard down to keep my feet from going through the holes while I was driving. But otherwise the car was okay, and whenever we found any free time, we traveled around the state and even went to Niagara Falls (the American side only, since we had no green cards yet and could not travel to Canada).

The biggest impact on the health of the car was Irena's practicing to pass her driving test. In Ohio, driving tests were serious business. The most difficult part of the test was parallel parking. I mastered it pretty well, and even now, sixteen years later, it pays good dividends, because I can park any car on any New York street, squeezing between two other cars with just a hair's breadth to spare. But for Irena, parallel parking was an ordeal. She failed that part of the test three times and kept practicing on deserted parking lots. I would make marks on the ground and she tried to maneuver the monstrous Electra between them. Since the car was about as long as half a city block, parking it demanded inhuman skill. I always wondered just what kind of streets these machines were designed for? They would have been perfect on the vast prairies of the wild West, but in city streets... Anyway, Irena finally passed her test, but the car's front end, the power steering system, and the transmission all took quite a beating, which very soon made itself known.

After one year, our landlord, a small but vicious woman, raised the rent. She did not like us, newcomers who had barely arrived in this country before they had a job, an apartment, and even a magnificent-looking black Buick Electra. Obviously, her own arrival in this county three decades earlier had been far more difficult and painful. The poor woman had known great tragedy in her life—she had lost her family, and she herself had survived Auschwitz, and I suspected it was there that she developed her callous nature.

She raised the rent so high that we were simply unable to afford it. I went to Dr. Plonsey, the head of my department at the university, and asked him if there was any way I could get a teaching job as an assistant professor, or just get any other kind of raise.

He smiled and said, "Are you kidding? This is a great university. People pay *us* money just for the privilege of working here, and you have the nerve to ask for a raise?"

Of course, he was joking and he went on to say that he would like to do that for me, but that just then there were no openings in the department. "People stay with us for a long time. We rarely have an opening and that's usually when somebody dies, but there's no knowing when that might happen. It could be tomorrow, it could be five years from now."

I told him that I could not help anyone die and really appreciated the great honor of working for Case. But my intelligent landlord also cherished my university-made money and wanted even more of it. Kindhearted Mike Neuman was able to beef up my annual salary by about $2,000 from one of his grants, but even that was not enough to support our more than modest lifestyle. In the end, I started looking for another job.

The first part of our lives had been spent in the Urals, which is really part of Western Siberia, a thousand miles from any sea. We had known there only forests, high hills, and small lakes. Now we badly wanted to live near the sea or ocean. I loved snorkeling, walking along the beach, swimming in the waves. Living on the coast was my as-yet-unfulfilled dream. Therefore I started sending copies of my resume to places on the two American coasts and nowhere in between. I decided that living somewhere either around New York City, by the Atlantic Ocean, or in California, Washington or Oregon, by the Pacific, would be ideal. I mailed out about a hundred letters and received several replies. All of them said, "No, thank you, you are overqualified for the position." I was naive enough to mention in my resume that I had a Ph.D., which apparently scared off a number of potential employers. If I were more modest and simply included my E.E. degree, no doubt I would have more easily found a job.

One evening, I got a telephone call from the owner of a small company in New Haven, Connecticut. The company made sophisticated electronic equipment for biological intracellular research. He was looking for a biomedical engineer with a strong background in analog circuits—a rare quality in that dawning age of microprocessors and digital systems. I was not only exactly the kind of engineer he needed, he also hoped to get me cheap, since I was a newcomer to this country. He invited me to come to New Haven for an interview, and I was modest enough not to ask him to pay for my airline ticket. I decided to drive to Connecticut, especially as my wife also wanted to go with me and see for herself what New Haven looked like and whether it made sense to give up our charming Cleveland for a small New England town, even if it was the home of Yale University.

In the middle of August 1979 we boarded our limo-like Electra and set out toward New York City, that being the most direct route to New Haven. The road was pleasant, and Irena sat in the back seat with Roman and read him fairy tales in Russian. Shortly after we passed Youngstown, we saw dark clouds. A storm was approaching and we were headed straight for it. Soon a torrential rain was pouring down in sheets over the fields, the road, and us, like a vertical river falling from the sky. The wipers danced like mad, but their efforts to clear the vision were futile. Driving became impossible, and we pulled over to the shoulder to wait until the rain let up. About thirty minutes later the storm passed and we continued on our way. We drove along for a while at a modest 45 miles an hour when suddenly we felt a tremendous shock and our huge car leapt into the air like a grasshopper. Apparently, we had struck a pothole which was filled with water and thus invisible. But most terrifying of all, I completely lost control of the car. The brakes were gone, I could not steer, and we skidded off the highway at full speed. The car fell into the ditch along the side of the road, flipped over and landed on its side. Miraculously, nobody was hurt and we managed to crawl out onto a grassy knoll. To our surprise, the car had no visible damage, apart from a few clumps of grass stuck in the front grill, like in a cow's mouth. In those days, American car makers build very sturdy and solid car bodies.

It was then, for the first time, that I experienced the fraternity and emphatic nature of American motorists. Several cars pulled over and people asked us whether we needed any help and offered to send us a mechanic and a tow truck. Indeed, half an hour later a tow truck arrived to pull us out of the ditch and tow us to the nearest repair station. There we learned that the impact with the pothole had knocked all the belts from the engine, which was why the car had spun out of control. The mechanic replaced the belts and told us that, as far as he could guess, everything else should be okay. At that time I already had my Master Charge card (this being, after a car, the second most important condition of becoming a real American), which I used to pay for the service, and we continued our trip, thanking God for having spared us

Somewhere in the mountains of Pennsylvania, just as our car had successfully climbed to the top of a ridge and we were starting our slow descent, I heard a strange rattling noise coming from the motor. The car began to shake and the rattle grew louder and louder until finally the motor let out several sounds like gun-shots, and we saw thick black smoke billow out from under the hood. I pulled over and killed the engine (which was dead already anyway)—a very wise move, as we then saw flames licking out of the front end of the car. I opened the hood and poured all our drinking water over the engine. That put the fire out, though smoke and steam still enshrouded our poor Buick. In a few minutes, however, the strong mountain wind blew the smoke away and our metal horse stood there breathless and cold. Looking at it, I realized that my first car had finally expired, and that no emergency medicine could revive its badly battered hulk.

We decided to set out on foot. Only one thing bothered me: what should we do with the car? Obviously, it still had some value, but on the other hand, getting it towed down from the mountains might well turn out to be fairly expensive. Besides, I would then certainly be late for my interview in New Haven, and I had no wish to press my luck. So I unscrewed the license plates, we packed our belongings into two bags we had in the trunk, I set Roman on my shoulders and we started walking down from the Appalachian mountains, abandoning our fond metal friend forever.

We cannot have presented a very cheery picture trudging along the shoulder of the freeway. We decided to try hitching a ride, and before long a young man in a pick-up truck pulled over and offered us a lift. I still cannot forgive myself for not asking him his name, because that kindhearted fellow drove us all the way to New Jersey, right to Elaine's house, and never took a penny from us. God bless him for his kindness and generosity. Ever since then, I myself have always tried to offer whatever help I could to motorists in need, though I know very little about cars.

Elaine took us to Grand Central Station in New York City, where we boarded a train for New Haven. Once there, I successfully passed the interview, received an offer of a job and negotiated a salary, and we even managed to find a lovely apartment in the nearby town of Branford. The owner of the company paid for our return airline tickets to Cleveland, and we returned home to get ready for our move to New England the following month.

When our friend Maria Palley asked Roman how the trip to Connecticut was, the four-year-old rascal said, "Bo-o-oring." What else did he want for excitement? An earthquake maybe? A war?

When we got back to Cleveland, I called our landlord and gave her one month's notice that we would be moving out, and we immediately began looking for a new car. I had some money saved from my moonlighting, enough for a down payment. Since we had no car to get around in, two of my colleagues from the university, Matt and Spiro, and the son of a friend, Alex Gelman, drove me about from one dealer to another. After about a week of intensive searching, we found a Toyota Corolla, a small but nimble station wagon. Since I had the offer of a job at a better salary, I was able to secure a loan from the bank, and within a few days we proudly brought home our sparklingly new car. My future employer had agreed to pay for the moving of our furniture, but we had to drive ourselves to Connecticut. After buying the car, we had only $100 left, which would cover food and gas on the trip to New Haven, but otherwise, we were not much better off financially than when we first arrived in Cleveland nearly two years before—except that now we had a new car.

I called our landlord again and told her that I had no money to pay the September rent and suggested she keep the security deposit as our last month's rent. She refused, saying that we had to pay the rent and that only after we moved out would the deposit be refunded. Legally, of course, she was right, but at this point it was not a matter of choosing to pay or not to pay. I just did not have any money left, plain and simple, and so we decided that the security deposit would have to be sufficient compensation, especially since we planned to move out well before the end of the month.

We planned our departure for September 17. The day before, a moving company picked up our furniture, while we prepared to load the rest of our belongings into our new station wagon and take off very early the next morning, for it was about a ten-hour journey to Connecticut.

It was around nine o'clock that night when we heard police sirens. My first thought was that Cleveland never rests. Then the howling of the sirens came closer to our house, and I thought that maybe some kind of accident had occurred in the neighborhood. I looked out the window and saw a SWAT team armed with machine-guns surrounding our building. What was going on?

There was a knock at the door. I opened it and saw a police officer with several gunmen in bullet-proof vests behind him. The captain asked if a Mr. J. Fraden lived here, and then handed me a sealed envelope.

"What is this?" I asked, totally shocked and puzzled.

"This is a court order. And I have my own orders."

I opened the envelope and tried to understand what it was all about. I had lived in America for less than two years and my English was still not that good, let alone sufficient for understanding the legal lingo of those papers. But I managed to figure out that my landlord had complained to a judge that I might leave the state without paying her the last month's rent. And the judge, believe it or not, had ordered my new car impounded and held until the case was resolved. The police officer told me that he had to tow my car away. I showed him the papers and he could scarcely believe it himself.

"I wasn't told what this was about," he said. "My orders were just to impound the car. We thought it was some kind of drug business or something. That's why we came fully prepared.

And all that for the lousy rent!? What a joke!" He was very disgusted and even baffled, but, nevertheless, they carefully searched my car, hooked it up, and towed it away.

I needed advice but had no one to ask. At that time I was in the middle of negotiations with a potential investor for one of my inventions. The contract was prepared by his lawyer since I could not afford one of my own, so he was the only attorney I knew. The next morning I called him and asked him what I should do. He told me to come by his office, and when I arrived, he took a look at the court order and shook his head.

"I don't know," he said, "this isn't my area of expertise. It looks like your landlord and the judge are good pals. As far as I know, in Ohio the law is on the side of a landlord. But you missed the important point—the hearing was set for nine this very morning, and it's now ten to nine. You'd better hurry over to the court house, or you will be in trouble."

I ran like hell to the train station, and less than half an hour later walked into the court room, but it was already too late. When I had not shown up at nine o'clock, the judge had ruled by default that I could get my car back only after I paid the rent, the towing and storage charges and the police fees, and that, in addition, I would not get my security deposit back. I was told all this by the court clerk. The worst thing was that to get my car back I had to pay $650 cash, and every additional day of storage would cost me another $50. I did not have the money and had no idea where to raise such a sum. The court clerk suggested that if I had a credit card, I might try to borrow some money on it. I went to the bank, but my Master Card balance was already nearly at its credit limit. Fortunately, the people at National City Bank were very kind and understanding (I have always had luck meeting kind people) and agreed to raise my credit limit so that I could borrow the cash on my card. All this took some two days, at which point I paid the money to the court, got a receipt and picked up my car from the storage garage.

It was the last day of our life in Cleveland. The next morning, with mixed feelings, we left the city for good. On the one hand, it was the birthplace of our American life. On the other hand, I was on the verge of a nervous breakdown after several sleepless nights and all the ordeals with my car. It is no surprise that I wanted to hit the road as soon as possible. We stuffed all our belongings into our recovered Toyota and, with no further difficulties, drove straight to New Haven.

For eleven years we lived in Hamden, a suburb of New Haven. I love New England. It is so picturesque and artistic. Mother Nature was very generous to that part of the country, especially during the fall when leaves of the trees and the bushes turn brilliant colors.

Except for the fall and the spring, however, the climate in New England is not all that great. Actually, the climate almost everywhere in America is far from pleasant. Few places can be proud of their weather. The some I know of that can are Vermont, Colorado, and San Diego. In many other states, it is hot and humid in summer and cold and damp in winter. In one sense this may be a blessing, because I have a theory that a bad climate is an essential condition for human creativity. Modern civilization is founded on the fruits of creative efforts by nations lying in regions of the world where the climate is not conducive to prolonged outdoor recreation. The technology, science, and arts that we enjoy today have come mainly from Europe, the British Isles, and the Scandinavian countries, whose climates are all far from ideal. Bad weather encourages one to stay inside and do something productive, rather than going to the beach or playing ball all day long. These pleasant activities would never have advance civilization to the degree it has gone. In warm regions where a person just needs to stretch out a hand to pluck a banana or an orange from a tree, there is no inspiration and no driving force to work hard for food and shelter. An easy life leads to stagnation, while one of hardship leads to inventiveness. But in 1979 we moved from Cleveland to New Haven, from one place with less than wonderful climate to another, where I expected to work, not play.

New England, thanks to its proximity to two great cities, New York and Boston, is full of art and history. Some say that American history is relatively brief, but New England can challenge many old European cities and nearly all of Russia. As soon as we relocated there, Irena auditioned for the New Haven Symphony and several East Coast chamber orchestras. Once again, she was practicing day and night and going to try-outs auditions, and in the end became a permanent member of the symphony. She had, however, a tremendous appetite for work, and the Connecticut orchestras, which did not offer full-time positions, were not nearly enough for her. Quite often she traveled to New York and freelanced there.

I had been hired as chief R&D engineer. My knowledge of analog electronics and sensor technology was put to perfect use in my new job. The beauty of working for such a small company was, I found, the complete absence of red tape. After testing and a design review by my colleagues, my engineering ideas were all quickly realized in production. That dynamic functionality helped to foster innovation and made a positive impact on sales. The corporation for which I worked was especially efficient in that sense, because Harry Fein, the owner of the company, was himself an excellent engineer and inventor. He had a ready appreciation of the benefits of a new development or improvement. In almost every instance, I would manage to convince him with my proposal, he would make his decision then and there, and—boom—it was on its way to the test lab and, after that, to the production floor. As an engineer and inventor I could hardly have dreamed of greater professional satisfaction.

While living in Russia, we had very little opportunity to travel. Of course, we visited many places in that huge country, but apart from a few exciting places in Siberia, the Black Sea coast, and a couple of beautiful cities like Moscow and Leningrad (now St. Petersburg), there was really nothing to see. There were also several western-type cities, such as Tallinn and Riga in the Baltic republics, but that was about it. The Russian terrain is quite flat, but picturesque. Travel abroad was virtually impossible, which is probably the reason why many immigrants just like us, having barely arrived in America, set off traveling to every imaginable place in the world. A desire to compensate for lost time would drive them to Europe, the Far East, Africa, South America, the Caribbean islands, Mexico. An *idée fixe* was planted forever into our brains: to go abroad is a great adventure and an incomparable source of excitement. Even now, after living in America for more than sixteen years and traveling all around the world, I still feel an inner thrill when I cross the border from, say, San Diego to Tijuana, which is by no means the most exciting of travel destinations. This sort of feeling is very typical for many immigrants from the former Soviet block.

When we arrived here, a refugee had to live in the United States continuously for at least two years before receiving a green card, and only after that could he or she apply for status as a permanent resident. The very next day after our two-year anniversary of living in America, we drove to Hartford, where we successfully passed an interview at the Immigration and Naturalization Service, and a few months later we received our green cards. Actually, we did not receive the actual cards, but simply a letter stating that the INS had granted us the cards, which would arrive within three months, and that meanwhile we could use the letter as a legal substitute. We could wait no longer—we had to go *abroad* and see a foreign country! The nearest country to Connecticut being Canada, we decided to drive to Montreal.

When we arrived at the border crossing, we just drove through to the Canadian post. That alone was a tremendous feeling: no one in America checked our documents, no one asked for exit visas, no one searched our luggage or our persons. We simply left America and went abroad. The American border guard merely waved at us.

Alas, it turned out that getting into Canada was not so easy. When the Canadian immigration officer asked about our nationality, I said that we were stateless and handed him our letter from the INS declaring that we were permanent U.S. residents. The officer stared at the letter as if he had never seen anything like it before. He then took it to his supervisor, who came out to our car and told us that they could not let us enter Canada with that strange document. I tried to convince him that it was a fully legitimate document, but he said, "No, you must turn back. Go to America, if they let you in, of course."

Oh, it was too funny! What if the American immigration officers refused to let us back in? Where were we going to live? On that narrow strip of land between America and Canada? Great!

But when we turned around and drove back to the same guard who had waved us through just few minutes earlier, he was very surprised.

"That was one quick trip to Canada!"

I explained to him that the Canadian immigration officers had not liked my papers. He looked at the letter and said that there was nothing wrong with the document and that it was a totally legal substitute for a green card.

"You folks wait here. I'm going to call my Canadian counterpart and tell him that your paper is okay. Don't worry, be happy."

That nice fellow went to the office, but five minutes later he came back quite confused.

"Those idiots don't understand. They're just stubborn and there's nothing I can do. Sorry, folks, for your bad luck. Go and spend your vacation somewhere around here, in Vermont."

We were disappointed but not discouraged. Canada has many doors. If this door was closed, we would try another one. I looked on the map and found another road to Canada. We turned east and in less than an hour came to a small building at another border crossing between two countries. It was in the middle of a cornfield, and no one was visible nearby. I did not feel comfortable just zooming through the border and knocked at the door of the little house. The door opened and a pleasant-looking old gentleman appeared on the threshold.

"Yes?" he said in French. "What can I do for you?"

I told him that we wanted to enter Canada and handed him my paper. He looked at it, and I guessed that he could not read English and did not understand what I wanted from him or why I had stopped at his post

"If you want to go to Canada, monsieur," he said, "just drive straight to Montreal, where there's a beautiful flower exposition. It is a charming and a must-see event. Bon voyage!"

We thanked that nice man and headed to Montreal, which was indeed one of the most charming cities I have ever seen before or since.

Two years later, when we already had our green cards (which actually were blue), we decided to take a vacation some place even more exotic than Canada. And what could be more exotic than the land of pirates, that is, one of the Caribbean islands? However, even with our green cards, we still needed to obtain visas, which turned out to be quite a difficult enterprise. First, we had to apply for an American travel document, which could take several months to arrive, then we had to travel to New York, to the consulate of the country we wanted to visit, and apply for a visa. In the end, we decided to avoid these hurdles and go to an American island, St. Thomas, one of the American Virgin Islands. Apparently, this would not require a visa or even our green cards, which our travel agent even suggested we leave behind, at home.

We flew to Charlotte Amalia, the main town on picturesque St. Thomas, and spent an exciting week there of swimming, snorkeling, and sightseeing. It was our first vacation with all four of us together: Irena, me, and the children, seven-year-old Roman and two-year-old Julia. It was especially piquant that our tropical holiday was during Christmas, when New York was cold and windy. But in the American Virgin Islands, it was everlasting summer and serenity. License plates in St. Thomas bear the motto "American Paradise." I thought that it was paradise indeed.

One day, we took a taxi to a remote area of the island. As we drove along a winding mountain road, the driver sighed deeply and said, "Oh, man, life is very difficult on the island."

I could not understand how this could be, and asked him, "What is so difficult about your life?"

"Because you have to work to live, that's why," said the angry driver.

I was surprised. That man truly thought that he lived in *paradise*, and the fact that he had to work there, like mortal people do on Earth, was very disturbing to him. I just shook my head. Well, he had to work in paradise—let him work and earn his pay. But we had come there for a vacation, and whatever it may have been, hell or paradise, we enjoyed our first vacation "abroad" as never before.

When our vacation was up and we stood in line at the airport to board the plane for New York, we noticed to our horror two immigration officers at the gate. We had thought that St. Thomas was America and left our green cards back in Connecticut! How were we going to prove to them that our home was in the United States? With no documents, we would be in deep trouble. I decided to tell a small lie. The only question they were asking the passengers was, "Where were you born?" I told Irena that we had to answer, "New York." Since virtually everyone in New York speaks with some kind of an accent, and the immigration officers themselves did not speak the King's English, I hoped that it would work. And it did work, perfectly, for Irena and me. But then one of them turned to Roman.

"What about you, young man? Where were you born?"

The little rascal proudly and loudly announced, "I was born in Russia!"

We stood there petrified, but the immigration officers laughed uproariously.

"How about that! His mom and dad were born in New York, and this little joker claims he was born where? You'll never guess! In Russia! Get into the plane, you little liar!"

When we had lived in America for five years, we submitted our documents for naturalization. I was very excited and waited impatiently for the interview and language examination. The interviewer was a rather dull and faceless government bureaucrat. He could clearly see that I spoke English well enough, but nevertheless he pointed first to a chair, then to the window, then to his nose, and asked me to say those words.

When I think back on Russian bureaucrats and compare them with their American counterparts, I think that I would rather prefer dealing with the Russians. The American bureaucrat is not human. He (or should I say *it*?) is a computer with a rigid program that has few subroutines. It's like a video game. You can win if you play by the rules and practice well. But if you try to be creative, or just cute, forget it. You will lose for sure. A Russian bureaucrat was old-fashioned and corrupt. The most complicated piece of equipment he ever saw was a mechanical typewriter. He knew very few rules or regulations. His sole concern, like that of any bureaucrat anywhere in the world, was to preserve his own status quo. So long as you avoided touching that sensitive nerve, he might turn out to be sympathetic and feel some compassion for you. Being still human, he might even help you. Moreover, if you knew how to grease his palm (not necessarily with cash, but perhaps just by returning the favor somehow), you almost certainly got what you wanted. But in America, you must know rules of the game when you deal with a computer.

With that pre-programmed interviewer, I learned a crucial lesson in dealing with government officials: *never volunteer any information.* Later, I heard a joke about the three Great American Lies. One of them is unprintable, another is "The check is on the mail," and the third is "I am from the government and I am here to help you."

Anyway, I obviously should have stuck to answering with a simple yes or no, but instead I tried to be entirely truthful and open. When he asked me if I had served in any foreign army, I should have just said no and left it at that, but instead I said, "No, but through the university I got my military training and the rank of a lieutenant of the reserve."

The fellow virtually jumped on me, and I realized that I was in trouble. Convincing him that it was just a mandatory part of my university curriculum was almost useless. Somehow I managed to assure him that I had never actually serve in the Soviet Army, because he finally told me that I had passed the interview successfully and now just had to wait for a court ceremony when a judge would pronounce me a US citizen. Some other Russians who were interviewed were not as lucky. They also volunteered information, mentioning that as teenagers they had been members of the Komsomol, the youth organization run by the Soviet Communist party. Virtually every young person had to join the Komsomol if he or she wanted to graduate from high school and go on to college. Such "membership" was an inevitable part of life in the Soviet Union, but to a US bureaucrat it was the same as being a member of a foreign Communist party—a condition which categorically precluded one from obtaining US citizenship.

One such aspiring guy had been filling out the application form where one question was *"Do you favor the overthrow of the United States government by force, subversion, or violence?"* Thinking it was a multiple-choice question, the poor fellow checked "violence" and lost his chance to become a citizen.

A few months later, on May 4, 1984, Irena, Roman and I were sitting in the back row of a large courtroom. Three-year-old Julia was also with us, but so far she was one of just a handful of American citizens in that tightly packed room. We were among nearly a hundred people who had come there to become new Americans. Next to us we saw a Caucasian couple, but with a baby with Oriental features. I asked them what they were doing there. They told us that they had adopted a little Vietnamese child and that now she was going to became an American girl.

It was a very affecting ceremony. The judge read the oath, which we repeated aloud, and then she pronounced us American citizens. It was one of the happiest moments of my life. I was deeply moved by the overwhelming feeling of becoming part of this great nation.

I remembered how sometimes on the streets of Moscow, near the U.S. embassy, I used to see real Americans. I would look at them as though they were Martians. To me, they were aliens from another planet, who did not realize what an incredible blessing they enjoyed in being Americans. They walked like ordinary people, spoke like ordinary people, smiled and laughed like us—and yet they were so different! They were Americans! What a proud name, what a title to bear! And now, I had become one of them. And I was happier than I had ever been before.

8
Moonlighting

> Anyone who goes to see a psychiatrist
> ought to have his head examined.
> -*Samuel Goldwyn*

In 1979, I was working as a postdoctoral research fellow at Case Western Reserve University in Cleveland. The job was great, but the monthly paychecks were quite small, even for the late seventies. Out of my salary and what money Irena made playing for the ballet and opera companies, we had a good many bills to pay: rent, utilities, gas, auto insurance, kindergarten for Roman, food, clothing, medical insurance, taxes, repaying the loan from Tolstoy Foundation which had helped us come to America, and everything else a normal American family routinely pays. For us, however, things were more difficult than for the average American family on a similar income, because we were just starting our new life in this country. In a very short space of time we had to acquire what most people accumulate over years: furniture, clothes for various seasons, a child's toys, technical books—you name it. Once the bills were paid, we had absolutely nothing left for savings. When we came to America, we had not been able to bring anything with us, and we had no one from whom to borrow money. On top of everything else, it was soon clear that we would need to be getting another car in the very near future. The old Buick Electra we had bought for $600 was rusted through from the roof to the wheels, and I did not expect it to hold up much longer.

Thus, I was very excited when my boss at the university asked me if I would like to do some consulting work for a medical doctor, a psychiatrist as it later turned out, by the name of Freeman. This Dr. Freeman had come up with some weird invention and wanted a prototype built. For this, he needed an electrical engineer who could moonlight for him. He offered to pay $15 an hour—a lot of money for me at the time. It was a fairly big project. Getting the prototype up and running could easily take a hundred hours and possibly longer. Actually, Dr. Freeman had already had a prototype designed for him by another engineer. For reasons I could only fathom much later, the engineer had quit, leaving the prototype half-finished. Dr. Freeman was looking for someone who could pick up the unfinished project and make the machine work.

The project was strictly an engineering job. There was nothing to research or invent—just do the electronic design as contracted, put the hardware together, make any necessary adjustments, do some testing, deliver the device, and collect my pay. It was that simple.

The work of an electronic hardware engineer is something like playing with tinker toys. You take standard building blocks (transistors, resistors, capacitors, integrated circuits and so on), arrange them in various ways, and there you are. You can make a radio, a robot, even some weird machine for detecting ghosts—wherever your imagination leads you. Or one could also liken it to the work of a writer—simply by putting the various words together, you can assemble a poem, a novel, or a memoir. You just have to do it right. In both activities, it helps to have some talent and know the rules of the game—how to connect all the parts together so that they look nice and work properly.

Now that microprocessors have become the most versatile building blocks, the art of electronic design has somewhat shifted into the hands of the programmer. In the old days, however, the hardware designer had to know a good many tricks to make a device efficient, economical and reliable. In this respect, my background and expertise were fairly solid, and I knew more than a few tricks of analog and digital electronic design. The machine I had to make was a medical device and I was a biomedical engineer with an electronic background—so I fitted the desired profile of the consultant the inventor was seeking.

Since microprocessors at that time were not yet readily available and handy personal computers were still several years away, the entire device was a hardware assembly which had to be permanently wired to perform a very specific function. This function supposed to be the detection of a patient's level of psychological excitation. Or something like that. But first, let me describe Dr. Freeman, the inventor.

It is a commonplace that over time the way you live or work has a strong influence on your appearance and behavior. Or perhaps the opposite is true, and people select their occupations and companionships to match their personalities and appearance. We have all seen dog owners who look exactly like their dogs. In happy families, spouses often tend look alike and behave in a similar way. People become part and parcel of their environments and their jobs. Many doctors distance themselves from their profession and behave normally. At least, they appear normal to the rest of us. I remember a joke about Freud, the father of a psychoanalysis, who was told by a young colleague, "Our profession is so difficult! It's torture to sit there for an entire day and listen to someone's troubles."

"But," answered Dr. Freud, "who listens?"

Dr. Freeman was a psychiatrist, and I suspect that he had indeed "listened" for a long time, because his profession had made a strong impression on his personality. A gray-haired man about fifty years old, he dressed in shabby clothes and dusty shoes, and his entire appearance radiated a neglect for the earthy realities of life. It was as if the constant interaction with his mentally agitated clients had created a compensating temperament in him—he moved slowly, spoke softly, smiled timidly. He was, without question, a pleasant and very serene fellow, and yet—very different.

I cannot say that he was demented in a formal sense, but his way of speaking and walking and his whole behavior suggested to me that his occupation had shifted him off track slightly. After all, he was an inventor—and isn't that in itself a sign of some kind of mental irregularity? He was a good and honest man, but all the same—he was an inventor and a psychiatrist! Quite a combination!

Dr. Freeman had managed to obtain a patent for his device (proving what a skilled patent lawyer can do for you!). The patent alone cost him about eight thousand dollars—not a small sum, especially for the mid seventies. And now he was prepared to spend even more, whatever it took to make his invention work. While the amount of money he had spent on the patent sounded tremendous to me, he could clearly afford the additional expense without putting any great strain on his budget. He told me that he came from a rich family and that he himself charged his patients $110 an hour. In comparison, my $15 rate sounded very modest indeed.

But back to his invention. His idea was to make an attachment for a telephone set. Like many self-employed doctors, Dr. Freeman wanted above all to see his patients as rarely as possible and to collect payments as often as possible. (Not too crazy, is it?) Instead of interviewing a patient face-to-face in the office, he wanted to talk to him on the telephone. The telephone gadget I designed for him was to have three colored lights, like a traffic signal: green, yellow, and red. The device was supposed to distinguish between the voices of the two talking parties. The doctor's voice would be completely blocked out, while the patient's voice was to be analyzed.

This analysis included measuring the pauses between the words uttered by the patient. Dr. Freeman's idea was that the longer the pauses between the words—that is, the longer the patient hesitated before saying another word—the more mentally disturbed he or she was. When a person spoke normally and his speech flowed smoothly, the doctor would see a green light on his telephone gadget, indicating that his interlocutor was not a good choice as a patient. If his speech was choppy, with ... long ... pauses ... between ... words, then the red light would come on and the doctor would know that the patient was a good candidate for throwing on the couch, or whatever psychiatrists do to their victims. And if the yellow light appeared, indicating that the pauses were somewhere in the middle, this would mean that the patient should be watched closely to determine in what direction his emotions were tending.

Moonlighting

I took over the project. Dr. Freeman and I discussed it daily, and I spent long hours working on it in the laboratory. A hundred years ago when people worked twelve hours a day, it was called slavery. Today, when someone works fourteen hours a day, it's called moonlighting.

I had not only to design the machine, but also physically to put it together myself, without the help of any machinist or technician. It therefore took a good deal of my time, and for a couple of months I was coming home quite late every night. Figuring out from the schematics what the previous designer had in mind was a big headache. As usually happens in such circumstances, I ended up tearing almost everything apart and designing my own version of the machine. To make a long story short, within about three months I completed the machine, tested it using myself and a tape recorder to simulate the other party, and then called Dr. Freeman to tell him that the project was finished.

He came over to the laboratory at once. I gave a demonstration, and he went off with the box to his office—obviously hoping to have a lot of fun with it. Several days later, however, he called me and said that the machine worked fine, but that I probably had to readjust the "craziness threshold," as he put it. It seemed the machine flashed a red light for virtually all his patients.

On one hand, he was delighted with this. But on the other hand, he suspected that it might be too good to be true. Should there not be at least a few who triggered, if not the green, at least the yellow light? Was it possible that all his telephone conversations were with mentally disturbed people? After all, he had spent time talking not only to his patients, but to various friends and relatives as well. Could it be that all of them were so far over the red line that they needed his professional intervention?

There were, as I saw it, just two possibilities—either the machine was flawed or the idea was flawed. I thought the machine was sound and Dr. Freeman thought that his idea was sound. In any event, he asked me to take a look at the device and see whether anything was wrong with it. Perhaps it really did just need to be readjusted or fine-tuned.

I took the box to the laboratory, opened it up, and very quickly found the problem. It was my mistake after all. In putting the apparatus together, I had switched the wiring for the doctor and the patient channels. The machine worked fine and its threshold levels needed no readjusting. However, since the channels were switched, instead of blocking out the doctor's voice, the circuit ignored the patient. Thus, the machine was detecting the verbal pauses in Dr. Freeman's own speech. And the way he usually talked was ... nearly ... the ... same ... pattern ... that ... triggered ... the ... device ... to ... turn ... on ... the ... red ... light.

The machine was perfectly sound, appropriately detecting the degree of its inventor's own craziness. He himself was the real cause for the red light! His invention really worked. Embarrassed, I decided not to confess to him that I had made a mistake. It was not that I was afraid of him, or felt bad about my error. Such things happen. Rather, I did not want Dr. Freeman thinking that it was he and not his patients who needed psychiatric help. He was a fine man and I had no wish to upset him. In any event, I did not feel that it was my duty to break the news to him about the actual cause for the red light—especially before he wrote me my check. I simply told him that the machine had not worked properly because of a defective transistor which I had replaced, and that now everything was fine.

It was a happy solution for both of us—he got the toy he wanted, and I got my pay.

Soon after that, I made a deposit on the first new car I had ever owned. It was 1979 Toyota Corolla. It sounds like a fairy tale now, but at the time the car cost just $5,000. And as I am writing this story today, in 1993, it still runs pretty well. Of course, I don't drive it anymore. About seven years ago I gave it to my father who had just arrived in America from Russia, and later he passed it on to my nephew Eugene. The boy is very happy with that great little car which I bought with my psychiatric money.

9
Up the Creek without a Paddle

> Experience is a good school.
> But the fees are high
> -Heinrich Heine

While working on biotelemetry systems and other medical devices at the university in Cleveland, I kept looking for something different. Not that I thought my research was not worth the effort. These projects were quite interesting and certainly had practical uses, though on a relatively small scale. I probably disappointed Mike Neuman, who, I believed, expected that I would take on some serious research project and produce useful scientific results which could be published in respected journals, for publication was the ultimate product of a research institution.

My interests, however, lay not in research, but rather in invention. I wanted to produce things that did not exist solely on paper, but could also be put into wider practical use. I was a trained scientist and my work relied heavily on science, especially physics. Yet I wished to create something more tangible than a piece of paper. Solving riddles posed by Mother Nature was not my cup of tea. The sciences study facts which already exist, outside and beyond us. Scientists investigate, they do not create. On the other hand, artists, architects, writers, inventors, and others in the creative professions produce things that have never existed before. I wanted to be like them, to make something new of my own—be it movies, paintings, or unusual electronic devices.

To choose my course in America was not difficult. I had changed my country, my friends, and my language. I spoke English with an accent and could hardly contemplate going back to what I had loved years ago—making movies or playing on the stage. Though I love these things, my opportunities there were long gone. I had no name and no real experience in those professions, and it was too late for me to start over from scratch. Clearly, an artistic occupation was out of the question for me in America. The only other profession I was reasonably good at was inventing electronic instruments, and so I decided to move in that direction.

The start of my life in America coincided with the Arab oil embargo and the tremendous gasoline shortage. There was, however, a widespread belief at the time that it was the American oil companies, rather than the Arabs, who were actually behind the crisis. People believed that the oil companies had orchestrated the shortage in order to send fuel prices—and thus their profits—skyrocketing.

Irena and I never used our car much. Usually I took a bus to the university. But for many Americans the gas shortage was quite a problem. They arrived at gas stations to find signs declaring "No gas," or else had to wait in endless lines to get just a few gallons. One rich man in Cleveland was so annoyed by all the aggravation that he bought a gas station just to avoid waiting in the lines when he needed to fill his tank. It was a great solution for him, but hardy a useful idea for the rest of us. A new type of crime had emerged—stealing gasoline from cars in parking lots. Some people were even killed in this struggle of nerves. Reports of such incidents triggered the entrepreneurial instinct in one bright man, who bought a surplus supply of gasoline tank caps with locks and keys from a warehouse for $1 a piece and within one week sold over 100,000 of them for $10 each, turning a very tidy profit.

It was late fall of 1978 when, one lunch hour, I was walking through the campus of Case Western Reserve University with my good friend Dr. Igor Palley. By way of mental exercise, we were racking our brains searching for possible ways to ease the energy crisis. Naturally, we had no illusions of making a discovery that might actually resolve the global energy problem, but even a small thing could help. Igor suggested that we come up with a more economical car which would not need as much gasoline as the cars of the day. His idea was to use a flywheel as an energy storage device. A heavy flywheel would rotate at a high speed, and when the car needed to accelerate, the wheel would be coupled to the transmission to propel the car forward. When the car needed to stop, the break pedal would direct the kinetic energy back from the car to the flywheel. The wheel would spin faster, thus storing more energy instead of allowing it to dissipate inside the car breaks. Thus, as one drove such a car, energy would be pumped back and forth from the flywheel to the car drive. A motor would be needed only to set the flywheel spinning initially and to compensate for friction loss.

Of course, the flywheel idea was not new, but we believed that the time was ripe to revive it and see if it might fly. There were a lot of problems related to this approach. One of them was safety. If the wheel ever broke, the pieces would truly fly off in all directions, like the bullets. Another problem was the gyro effect of such a fast-spinning mass, which would make it very difficult to steer the car. One option, of course, would be to use two counter-rotating disks to cancel that effect. But still, we saw the tremendous complexity of the task, especially given that neither Igor nor I was an automotive engineer. Our expertise clearly would not be sufficient for resolving the many difficulties. After bouncing the idea back and forth for few days, we decided that we could see no quick and easy solution to the problem and let it rest. There had to be plenty of other ways to save energy.

One evening I was working late at the laboratory, and Igor was late with his work as well. Finally, he called me and suggested that it was time to go home. The buildings where our labs were located faced each other, and we met on the street to walk to the bus stop together, as we often did. It was dark outside and Igor said, "Look at all those windows. Do you believe that everyone at the school works that late?" Indeed, most of the windows in the university buildings were brightly lit. Obviously, some people might be working late, but surely there could not be that many on a Friday night. People were just walking out without remembering to turn off the lights. Or perhaps the janitors were turning the lights on and leaving the rooms illuminated until the following morning, or, if it was a Friday, until the following week.

"That is a tremendous waste of energy," said Igor. "Is there any way we could make people not forget to turn off the lights? Isn't it shame to waste so much electricity?"

We reached the bus stop, boarded the bus and rode home. The bus was running along Euclid Avenue, the same road where at the intersection with 105th Street on August 5, 1914 the World first traffic light was installed. Igor's stop was first, while I had to ride on for another ten minutes. There were several traffic lights on the way home, and the bus had to stop frequently for red lights. Just recently, while speculating about flywheel cars, we had done some calculations to estimate how much energy is lost when a car or bus accelerates and stops. This made me think about those traffic lights which were forcing us to stop so often for no reason, since there were no other cars at the intersections. We just had to stop and wait for a green light and then proceed on again. Start-stop-start-stop... The loss of energy was colossal. Why stop and wait at an intersection, when there were no cars in the cross street waiting to go? In large cities with heavy traffic on long streets or avenues, there were "green waves" of traffic light controls. In New York City, for example, green lights along the avenues were turned on sequentially with six-second delays between the intersections to minimize stoppage of traffic. But what about smaller towns, or side streets in cities? It was obvious that this simple alternation of green and red traffic signals was causing a significant loss of energy and time. These two problems ran through my brain for some weeks: how to turn off lights in unoccupied rooms and how to put some intelligence into traffic signals?

I thought that it would be great to put some kind of "servant" in every room and at every intersection. Such a robot would have a simple yet quite boring job: watch to see whether a room was occupied, or whether any cars were approaching an intersection. In the old days, this would have been a good way to reduce unemployment. Today, the solution had to be of a more high-tech nature. What was really needed was a sensor which could detect people or cars and then signal a switch box either to turn the lights off at a certain point, or to change the color of a traffic light. Thus, I had to come up with a device that was sensitive to the presence of large objects, such as people or cars. Of course, there were many ways to do this. I could use microwaves, ultrasonics, laser beams—just to name a few. Yet all of these traditional ways of detection were too complex and expensive, whereas I needed a more simple and inexpensive solution.

To begin with, I analyzed the situations in which such a detector would need to operate. If a room was occupied, the people there were probably doing something—walking about, standing up, sitting down, watching TV, reading a book, scratching their noses, or making various other large or small body movements. Similarly, any car approaching an intersection was obviously in motion. A simple detector could not be designed to recognize either the shape or the nature of objects without being prohibitively expensive. In short, the only property common to both people and vehicles was motion. To be recognizable they had to be moving. This was the key observation. I decided to figure out how to make a sensor which would be responsive to the movement of objects in its vicinity.

Having studied human physiology, I knew that Mother Nature makes such detection possible in a very economical way. Take, for example, a frog. Its eye cannot see anything which is static, but once a potential prey moves, the frog's eye sends a signal to the brain and the frog acts accordingly. The frog's eye does not recognize the object, it detects only a change in the image contrast. This makes everything so much simpler. So I decided to make an electronic eye, or more accurately, a detector, which could see the contrast variations associated with motion.

Building on my knowledge of photography, I came up with the idea of combining a focusing lens and a light sensitive detector to make a contrast sensor, similar to that of a primitive eye. But unlike a camera, the motion detector had to have a light sensitive detector instead of film. To sense changes in image contrast, a grid would have to be positioned between the lens and the sensor. The lens would create an image on the surface of the sensor, but that image would be broken by the grid into many fragments. When an object moved, its image would cross the opaque and transparent sections, thus varying the amount of light reaching the sensor—in other words, the varying contrast would be converted into varying light intensity. In reality, I did not even need such a grid, as light sensitive resistors already existed with a serpentine grid-like shape on their sensing surface.

The problem was with the lens. I had to make a very inexpensive motion detector, and a glass or plastic lens could be a limiting factor. Then I recalled my childhood experiments with cameras. I knew that the simplest (read: inexpensive) lens that you can make is a pinhole lens. In effect, such a lens is just a small hole in a sheet of opaque paper or metal foil. It has an infinite focal depth and thus does not require focusing, as a glass or plastic lens does. So I took a small piece of aluminum foil, punched a tiny hole in it with a sewing needle, and placed it in front of a photoresistor which I bought at Radio Shack for 25¢. Then I put together a simple electronic circuit with an amplifier and walked in front of my newly invented motion detector.

The effect was great. The electronic circuit produced a strong deflection in the output voltage in response to my walking in front of the detector. After some adjustments, nearly a dozen different prototypes, and about a month of extensive testing at home and outside, I put together a demonstration device. My motion detector functioned only when it was light. Since it contained a light-sensitive element, it was not operational in darkness. However, that should not be a problem, since my goal was to turn lights off, not on. In a room with moderate illumination, the detector could "see" people's movement over a range of up to 30 feet, and on a street, it could "see" moving cars during both day and night (because at night cars would have their headlights on).

It was more difficult to devise a reliable electronic circuit for the light switch. The problem lay in the need to connect the motion detector and the switch in a series with the electric lights—that is, a tiny amount of electric power needed to be diverted from the lighting fixture to supply power to the motion detector. In the end, I designed such a circuit which turned out to be quite unique and patentable. That completed the entire design, and only at that point did I start thinking, "What's next?" I had never before been engaged in any business venture, nor did I really know what one did with an invention in America. I understood, however, that in order to proceed further, I would need a great deal of money for all the various steps of the enterprise, and that the first step was applying for a patent.

Dr. Freeman, for whom I had developed a prototype of his medical invention, told me that his patent had cost him $8,000. And that was only for the legal fees—how much money would I need to place my invention in actual rooms or at street intersections? I had no money of my own. I had been living in this country for just a year and was still poor as a church mouse. It was clear to me that the first thing I needed to do was somehow to raise funds for my enterprise. To this end, I began talking to people about my invention. Perhaps there was some risk in doing that without first filing for a patent, but what the heck—if you keep it to yourself, then you lose for sure.

A friend of mine, Simon Gelman, M.D., introduced me to another former compatriot, Alex Rutstein. His wife was a professor of music at Oberlin College, while he was a power engineer and had his own small consulting business specializing in energy management. Alex was tall, good looking man with gray hair and aristocratic manners. He spoke calmly and acted cautiously. His business was in good standing, and he had several patents of his own and definitely knew better than I what should be done next. It looked to me like a perfect match—what else was my invention but an energy management device? I told Alex about the motion sensor and asked whether he would like to take over the business side of the enterprise while I concentrated on the technical aspects. He agreed and we began devising a plan. It was quite clear to us that we would never be able to raise nearly enough money for the development, manufacture and marketing of our products. All that would require at least a million dollars and possibly much more. Who would risk entrusting such funds to two fellows with no track record and strange accents? Thus, we decided to do what the majority of inventors, like myself, attempt to do—find a company which was already operating in a similar market and license the rights to the inventions to it.

However, if you want to sell something, first you must own it. Therefore, we needed to patent the motion detector. We both had substantial experience working with patents, but never in this country, and so the need for professional help was obvious. But how were we to finance even that smaller portion of our start-up business? Just this initial stage alone would require at least ten thousand dollars, while the total cost of marketing the idea to a large corporation would probably run to somewhere between thirty and fifty thousand dollars. We began looking for investors to back our work. Since we were talking about only tens of thousands, not millions, the possibility of finding such an investor seemed more realistic.

In our search for an investor we met quite a few people who were in a position to spare a few bucks. Three of them, however, I remember very well, for they were rather typical, even quintessential, of the rest. Here they are.

Investor No. 1

The only really well-to-do man I knew at the time was the psychiatrist Dr. Freeman. I called him and described my invention to him. He said that he would like to see the prototype and we set up an appointment at my home. Before his arrival, I tested my demo prototype and it worked great. It was housed in a black plastic box with a power cord and an electric outlet on one side, into which I plugged a floor lamp. It had a tiny red light which flashed every time the sensor detected movement in its vicinity. There was also button on the top for turning the lamp on.

Dr. Freeman arrived in an old and rusty Chevy which burned more oil than gas. His clothes were worn and his shoes had never seen polish. Perhaps this was his way of letting me know that he did not waste his money on nothing, or, perhaps, I thought, it was how a man of great mental abilities should present himself to look cool. I invited him to take a seat on the couch, explained the need for saving electrical energy, and pushed the button on my prototype. The floor lamp lit up and the red light blinked. I walked back and forth about the room, waved my hands, sat down in a chair, stood up. Every time the red light blinked. The motion detector worked as expected. Then my guest and I went into the kitchen and waited there for about five minutes. When we came back to the living room, the floor lamp was off. Dr. Freeman liked what he saw. He said that he needed to think it and left without even asking me what I was looking for in terms of investment.

After that first visit, he visited my house at a least a dozen times in the course of the next three months, sitting on the couch in the living room, waving his hand and smiling at the blinking red light. He really liked the invention. One evening I told him that I was selling the equity in my invention. The price was 1% of whatever royalties we received for every thousand dollars of seed money, up to 50%, or fifty thousand dollars. Dr. Freeman just said that he needed to think about it some more and kept coming to my apartment, sitting on the couch and waving his hands, but never giving me a yes or a no.

Meanwhile, Alex, who was conducting his own search for an investor, would curiously ask me from time to time where things stood with Dr. Freeman. Finally, I called the indecisive psychiatrist and suggested that Alex and I meet him for lunch to talk about our proposal.

At the restaurant, we waited some forty minutes for Dr. Freeman. He was late as usual. He arrived in his familiar worn slacks and dusty shoes. We ordered salad and coffee, and once again presented our proposal. Instead of responding, he just sighed and uttered a few nebulous sounds. We asked him the same question in many different ways, but it was absolutely impossible to milk any meaningful response out of him. Finally, I lost my temper.

"Dr. Freeman," I said, "I see that you have difficulty in pronouncing certain English words. If you excuse my accent, let me say one of them for you."

He sighed again and said, "Well, if you want to..."

"Okay," I said. "You want to pronounce word NO. Correct?"

Dr. Freeman gave a feeble smile, nodded, and mumbled something that sounded like, "Well, you said it..." That was the end of our conversation. He left without even bothering to pick up the tab.

This was the first time I learned about a deficiency in the American character—a pathological inability to utter the simple word *no*. Later in my life, I faced that deficiency many times with frustration and despair. Americans are very polite and kind people—I even think that they are the nicest people in the world. No doubt they do not wish to offend anyone by rejecting them, and instead of saying *no*, they make up endless excuses, delays, allegoric expressions, etc. It took me many years to develop an ability to recognize such "diplomacy." Frankly, I hate diplomacy, and while a totally civil person, I admire a certain military directness.

There is a joke of the circular type about the profound differences between people, depending on the directness of their responses to questions. It goes like this:

> If a soldier says *yes*, it means *yes*.
>
> If a soldier says *no*, it means *no*.
>
> If a soldier says *maybe*, he is no solder.
>
> > If a diplomat says *yes*, it means *maybe*.
> >
> > If a diplomat says *maybe*, it means *no*.
> >
> > If a diplomat says *no*, he is no diplomat.
> >
> > > If a lady says *no*, it means *maybe*.
> > >
> > > If a lady says *maybe*, it means *yes*.
> > >
> > > If a lady says *yes*, she is no lady.

Leaving the lady out of this discussion, I hate the diplomat and like the directness of the solder, where *yes* means *yes* and *no* means *no*. Isn't it great? That's how I try to behave myself. I always say plainly what I want and what I mean. It saves a lot of time, money, and nerves. Many people, however, prefer the language of salons and the United Nations, which talks a lot and does little.

Anyway, Dr. Freeman was just a waste of our time, and we had no choice but to keep looking for a more decisive investor with no speech deficiencies.

Investor No. 2

One day, Alex called me and said that he made an appointment with a very rich man who wanted us to present our proposal to him. This man had a great name for an investor—Jack Gold. We met him in his spacious office in suburban Cleveland. Mr. Gold was a short old-timer with a wrinkled face. The walls of his office were covered from floor to ceiling with photographs of his parents in the nineteenth century, of himself in various shapes, hairstyles, and surroundings, of his children, grandchildren and great-grandchildren, dogs, cats, and other happily smiling relatives in all stages of evolution. He was in real estate or oil or some other profitable business (by the standards of the time).

With his permission, I used his desk lamp for a demonstration. Mr. Gold was very impressed with the prototype. He quickly grasped the idea of using a motion detector for energy conservation, and I, in turn, was impressed with his business-like approach.

Popping a hard candy under his tongue, he said, "Fellas, I love it. I've never seen anything that simple and effective. I think we're gonna make a deal. This is my plan. Listen carefully. Tomorrow morning I'm going to Florida. It's getting pretty cold out here, in Cleveland. I'm an old man, you see. I need to fry my bones under the southern sun. But I'll be back in just two weeks, so let's set up another meeting right now and we'll waste no time in making it fly. See you, fellas! I lo-o-ve it!"

During the next two weeks, while the charming Mr. Gold charred his old bones beneath the Florida sun, we worked out a business plan which included the patenting of both the light switch and the traffic light detectors, a search for licensees for the products, and my personal participation in the development process. We wanted to be ready to answer any questions that might arise during our negotiations with our investor-to-be.

We arrived at his office five minutes before our scheduled appointment and his cheerful secretary immediately ushered us in. Mr. Gold had changed noticeably since we had seen him two weeks before. His face and his bald head shone with a fresh brownish tan and his entire appearance radiated a total satisfaction with life. But when he began speaking, we did not like what we heard.

"Well, guys, let's get right to the heart of the subject. Tell me what are you offering me in return for my fifty-grand or so and I'll tell you what I think about all this commotion".

Alex, in his imposing and calm manner, told our sun-charred host that if the plan worked as we hoped, Mr. Gold should expect to make at least a million dollars after about two years, which was about a twenty-fold return on his investment.

Mr. Gold popped a hard candy under his tongue and said, "Look, fellas, I am seventy-three years old, you see. I don't know exactly how much, but I already have something like seventy million dollars in cash and assets. I am an old man, fellas. Tell me frankly, do I need another million? Shrouds have no pockets, ha-ha-ha-ha!"

"Good-by, Mr. Gold, rest in peace," was all we could mumble as we left his office. At least he was decisive.

Investor No. 3

Our search for an investor went on another few months until finally, in March of 1979, Alex told me that he had just talked to a rich mattress manufacturer by the name of Henry Goodman (what a great name, I thought: Good-man—so much better than Bad-man, or even Mr. Gold, for that matter). Apparently, this good man was interested in talking to us about our proposal. We met with him and he seemed like a very reasonable man. He took nothing we told him for granted, which is a wise way for an investor to go when he is talking to an inventor. He hired his own consultants, asked the opinions of a couple of patent attorneys and marketing analysts, and after several weeks of his own investigation and reflection, he told us that he was willing to consider the light switch, but that the traffic light was too risky. He explained that, according to his information, the manufacture and installation of traffic lights was a highly politicized business, in which having the right connections and knowing the right people were of paramount importance. He even mentioned something about Mafia involvement. Without the ability to pull the right strings, he said, we might take forever just getting the traffic light business off the ground, and still be at great risk of failure. Of course, he added, we might be able to fix up an intersection or two, and then try to sell the idea to some small town, just for starters, but even then, the whole thing would be too difficult and slow. We should do the one thing first, he said, and the rest we could consider later.

He insisted that we concentrate our efforts on a motion-controlled light switch, for which he was willing to provide financing of up to fifty thousand dollars. It seemed a shame to abandon the traffic light, but we had little choice. I gave up that invention and never did any further development on it. Now, fifteen years later, I see similar traffic signal controllers installed at many intersections all around the country, though they use a different kind of sensor, one which is imbedded in the pavement rather than installed directly in the traffic lights. We could have done the same thing a decade sooner for a much lower cost, and I think it was a lost opportunity for all of us.

We set up a limited partnership with Alex as a General Partner. (In the seventies, it made sense from a tax standpoint to organize such partnerships. This is no longer the case.) We divided the shares into three equal parts, with Alex's portion being further diluted as he wanted to engage a marketing man as his assistant. It was my mistake to agree to let Mr. Goodman receive a third of the stock without making those shares contingent upon his investment. But he was such a nice man that I did not want to offend him with even a hint of mistrust.

I hoped that the partnership would be able to "reduce to practice" not only the light switch, but also several other of my inventions of a medical nature. Thus we called the partnership "Bio-Optical Sensors, Ltd.," and Mr. Goodman loaned us his lawyer to help in setting up the partnership. He told us that his further participation and investment had to be organized in proper legal fashion and that it would take a few more weeks for his lawyer to draft a formal agreement between Mr. Goodman, the partnership, and me. He put down some initial money and we hired a patent law firm which did a thorough patent search. Such a search is a very useful step in product development, and a rare case when money is really well spent. To my great surprise, the search revealed that the principle of such a motion detector, which I had thought was my original idea, had first been patented in 1936, and that at least a dozen other patents for further modifications of the original idea had been issued since then. In truth, there is nothing new under the sun.

However, even though the general principle was already patented, there were many ways of making it work. This was why so many other patents concerning the same idea had already been issued. My design with its solid-state photoresistor, pin-hole lens and unique electrical power circuit was still quite original and, as Mr. Granger, the patent lawyer, told us, patentable, and the application had to be filed right away. Meanwhile, Alex and I kept calling Mr. Goodman to remind him that we were still waiting for his investment and the legal papers to formally wrap up the deal. In most cases, he was not available to talk to us, but when he was, he assured us that everything was under control and that we just had to be more patient.

During this time, I was looking for a job. The summer had arrived and we still had neither an agreement with Mr. Goodman nor the money to start selling the inventions, yet he already owned one third of the partnership. It was becoming more and more difficult to get him on the phone, and I started calling his lawyer to ask when the agreement would be ready for signing.

One evening, I received a call from Connecticut. It was an invitation for a job interview. Irena, Roman and I made a trip to New Haven. I successfully passed the interview and was offered a job. We decided to move to New England. We returned to Cleveland and began our preparations for the relocation. The final moving day was set for September 17, and that day was rapidly approaching. However, Mr. Goodman's lawyer still was not ready, and Mr. Goodman himself was still unavailable to talk to either Alex or me. I

Up the Creek without a Paddle

began to wonder whether this was the same syndrome we had seen with Dr. Freeman—the inability to pronounce the simple English word *no*. Did Mr. Goodman want the deal? If not, why not just say so? And if he did still want it, what game he was playing?

On September 10, I called his lawyer one last time and let him know that I was about to move to Connecticut for good, and that if the agreement was not signed before my departure, the deal was off. He called me back at the end of the day on September 15 and said that the documents were ready and that Alex and I should come downtown to his office at once.

It was about 8 PM when the lawyer, Alex and I gathered in a spacious office with a lake-front view. The lawyer put three sets of fairly thick documents on the table, opened them to the last pages, took out a pen from his pocket and pointed with the pen to the spaces where he wanted us to sign.

"Excuse me", I said, "but where is Mr. Goodman? How we can sign without him?"

"Oh, don't worry about that. He is a busy man, he will sign later on. It's perfectly okay. Just sign here and here."

"How we can sign without even reading the document?" asked Alex.

"Well, it's pretty late now. But you are right, take a few minutes, look it over. I'll wait, no problem."

Alex and I tried to read the document. At the time, I had been living in America for just twenty months and my English was still fairly shaky. Besides, this was the first time I had seen any American legal papers. Frankly, I do not think that many native speakers are able to read and understand what lawyers write. I have always suspected that doctors write their prescriptions in such terrible handwriting to conceal their simplicity and create some mystery about the medical profession. In the old days, they used Latin. Nowadays, however, when generic drugs and trade names have taken the place of Latin, unreadable handwriting has become the modern medical cryptography. Similarly, lawyers have invented their own pigeon language to hide the simplicity of meaning and charge amply for the translation.

Anyway, after ten or so minutes of struggling with this legalese, we realized that it would take a good deal longer to dig through the complex verbiage and the web of puzzling clauses. Alex said, "You are right, it's getting late and we're all tired. It's better if we take the documents with us, talk to our legal advisor, and get back to you in a few days."

"Wait a minute!" The lawyer leapt up from his chair. "I thought you were running out of time. Believe me, it's just a formality, a very standard straightforward document. Let me go over it very quickly with you."

"We appreciate that," said Alex, "but we'd better take our time. We need to understand it fully and we want to talk it over among ourselves."

The lawyer's face turned red. "Do you realize that your refusal to sign it today may appear somewhat insulting to Mr. Goodman?"

"Not at all. Mr. Goodman is a businessman of great experience and understands that an important document can never be signed on the spot." With that, we picked up our copies and walked out of the law office, leaving the disappointed lawyer behind. Alex had his copy and I took mine. Few days later I moved away from Cleveland, driving our brand new Toyota to New Haven. I have never seen Mr. Goodman again.

Shortly after arriving in Connecticut, I began reading over the agreement. Some of the clauses I understood quite well and others I grasped at least vaguely, but most of them were quite murky for me. A dictionary was little help. A word-by-word translation was not sufficient. I had to know the deeper meanings of all those strange-sounding expressions and their possible implications and consequences. Unlike me, Alex decided not to beat his brains out over the thing and did it right—he hired a lawyer who deciphered the meaning of the document for him. Once Alex understood it completely, he called me and we went through it together, paragraph by paragraph. The further we went, the more shocked I became. Only then did I realize what kind of game Mr. Goodman had been playing with us.

Some people have such a sensitive and brittle conscience that whenever they act shamefully or perform some foul deed, they modestly hide their faces, or at least blush. Apparently, Henry Goodman was just such a bashful man, for he had avoided all contact with me during my last several weeks in Cleveland, not even showing up for the document signing. The agreement his lawyer had prepared at his request was quite a nasty piece of paper, to say the least.

Among its many tremendously restrictive clauses, the document stated that his investment would be in the form of a subordinate loan, which was to be repaid from any income of the partnership, before either Alex or I could see a single penny. In addition, he would maintain his one third of the partnership shares, meaning that after he got his money back, he would still be receiving the same royalties as I. But the most shocking thing was that the document contained a virtual provision for slavery. It stated that whatever I had invented in the past or whatever I might invent in the future would be assigned to him, Mr. Goodman, and that he would thenceforth be the sole owner of all my inventions. Now, I know that such a provision would never stand up in court were I to challenge it later, but who wanted to go to court? Naturally, I was very angry, and I told Alex that under no circumstances would I sign such a document. Alex agreed and said that he would communicate this to Mr. Goodman and ask him to change the terms of the contract, otherwise we would go elsewhere.

It took about a month for a revised version of the contract to come back, but the only thing our investor agreed to remove was the slavery provision. He still got back whatever he invested, and only then would we be entitled to collect money from the remaining royalties. In addition, he would receive his one third of the royalties. We had already spent over a year looking for an investor and did not want to waste any more time. Deciding that this was better than nothing, we signed the agreement and set to work. Eventually, Mr. Goodman signed the contact as well, but he was so angry about my refusal to subjugate myself to slavery that he never returned any of my calls after that and never even responded to my letters and postcards. At the conclusion of the enterprise, he not only got his money back, but made a very tidy profit, though he probably still thinks that I was an ungrateful scoundrel.

Alex began contacting a number of companies engaged in the business of making lighting fixtures. He approached General Electric, Westinghouse, Nutone and many, many others. We hired an industrial designer and a model shop to make an attractive-looking demonstration prototype. We tested it thoroughly under various conditions. I prepared a detailed description, drawings, circuit diagrams, a parts list, and other technical documents a potential licensee might require in order to evaluate our proposal.

My partner worked very hard to make connections with anyone who might have an interest in making and selling our light switch. Alas, my personality and style were quite different from his. I was a fighter and a risk-taker, while Alex was a very cautious man. Many times he told me that, as general partner, he would be responsible if anything went wrong, so his strategy was to avoid even the most remotely hypothetical danger, and that only then would he see to the business of selling a license for my invention. He insisted that before we showed the prototype to someone, that party must sign a disclosure agreement protecting us. He wanted to have that protection in case anyone at such a company attempted to steal our idea.

For the most part, large companies tend to avoid signing any type of binding document. They are rightly afraid of the prospect of some sinister outsider suing them if they decline to buy his patent but then start producing something similar, even if it is an article or process they have developed independently. The disclosure agreement which Alex wanted everyone to sign became a real stumbling block. Nearly every company flatly refused to sign the agreement prepared by our lawyer, and Alex, in his turn, refused to disclose any technical information to them. It proved to be a Catch-22 with no apparent solution. Time and again I argued with him, insisting that we had to take a chance, especially after we filed a patent application. The patent pending was already considerable protection. Yet the business side of our enterprise was run by Alex, and I could not force him down what he thought was a dangerous road.

Time was marching on, then running, then flying. Before I knew it, I had been living in Connecticut for more than two years, but still no progress had been made in selling a license for the light switch. This lost time was the most risky part of our business. You cannot sit on your invention and wait. There are plenty of smart people around, and the idea you thought was so original and efficient may very quickly become obsolete.

Finally, Alex was able to reach an agreement with a New Jersey company which was one of the major suppliers of lighting fixtures. He signed some kind of secrecy agreement with them and asked me to ship my prototype to the head of their engineering department.

Within a week, Alex got the prototype back, completely broken and charred. Apparently, some idiot in the company's engineering department had plugged the switch directly into an electrical outlet. Our wall switch with its motion detector was still a switch, not very different from any other electric switch. And, as a switch, it had to be connected only in a series with an electric lamp. If you plugged it directly into an a.c. outlet, you had a short-circuit connection and—boom! And boom is exactly what happened, burning the thing up. I had to build another prototype and send it to the company, this time with a detailed letter explaining the difference between an electrical outlet and a light switch. (It certainly seemed a bit pathetic to have to explain such things to the engineer of a company which made electrical appliances.)

Again, we waited, and waited, and waited. Four months later, Alex received a letter from the vice-president of the company informing us that they had decided not to acquire the license for the motion switch. Their decision was based on the recommendation of the engineering department, which he attached to the letter.

The explanation the test engineer gave for rejecting the detector was patently ridiculous, if not grotesque. Apparently, it was the same not-too-bright guy who had plugged the motion switch into an electrical outlet. He had tested our prototype and found that it worked well under most conditions. Yet it was less reliable at detecting movement when a black man in black clothes was walking in front of a black wall. Under such circumstances the motion detector failed to operate reliably. I could not believe my eyes! So what? The detector was not a security or life-saving device—it was just a light switch, for heaven's sake! If you were worried about such conditions, the solution was not to install it in a room with black walls (and how often do you find such rooms anyway?). And even if it missed that unfortunate black man in his black outfit, the light would simply turn off a few minutes sooner. Big deal! Anyway, that was the excuse for rejecting our proposal.

Later, I learned the real reason why inventions offered by outsiders are usually turned down by the technical departments of large companies. Many technical departments suffer from the NIH syndrome, which stands for *Not Invented Here*. Think about it. Engineers at a company are paid to do their jobs, which often include developing new products. One day, an outsider comes in with something that they themselves might reasonably have been expected to come up with as part of their job. Partly from jealousy, but mostly from a self-preservation instinct, the engineers find any possible excuse to report to their superiors that the proposal does not hold water. How many good inventions never found their ways to consumers because of that NIH syndrome?

Alex continued to search for someone who would at least agree to talk to us. Another year passed. And another. We made a few more presentations, but none of them materialized into a licensing agreement. We had already spent about $20,000 of Mr. Goodman's money and saw no light at the end of the tunnel.

Of course, the competition was not sleeping. My motion detector operated in the visible spectrum of light, but a more efficient sensor could be designed with an infrared detector able to perceive the heat waves emanating from a warm human body. Such a detector could operate in a total darkness and not only switch a light off, but also detect someone coming into the room and turn the light on. At that time, such detectors already existed, but they were quite bulky and expensive. This meant that, in terms of a cost-effective light switch, my approach was still the best. Nevertheless, the prices for infrared sensors were beginning to come down, and I suspected that in a couple of years we would have nothing to sell. I was very discouraged and frustrated to see an apparently good invention become obsolete.

Realizing that we had to do something more radical, I mentioned our problem to one of the wheeler-dealers I had met at Timex, which I joined in 1981. This gentleman, whose name was John McManus, was a very energetic middleman with many Far East connections. He said that, for a piece of the action, he could arrange a reasonable deal for us with an overseas company. I called Alex and convinced him that we should take a chance with John, even if such a deal resulted in a further dilution of our shares. Alex's philosophy was quite pragmatic—better to have 1% of a million, than 100% of nothing—and he agreed readily. We signed a formal agreement with John, and a short time later he introduced us to Chuck Wheatley, the American rep of a Hong Kong-based company named Dimerco. This is how such middlemen make their money—they spend a few minutes on a telephone call, connect you with the right party, and then collect their dough. No wonder so few people want to produce anything and so many just want to make connections.

Since Dimerco was not an American company and was not as paranoid about being sued, they had no qualms about signing a disclosure agreement. The company's president gave Chuck instructions to negotiate a mutually acceptable deal with us. Alex and I flew to Cupertino, California, where Dimerco had its U.S. offices. With both parties eager to make a deal without wasting a lot of time, we reached an agreement in just one day, shook hands on it, and told our lawyers to draft a formal agreement. At last, we had licensed our patent to a company that sincerely wanted to bring it to the market. Five long years had passed since I built the first prototype of my motion detector.

During the next twelve months, Dimerco fabricated a production prototype and then came to the conclusion that it did not have enough muscle to enter the U.S. market and sell the switch to the American public. Once again, the fate of my invention grew murky. Japanese, Hong Kong, Philippines and other Far East-based companies love to own rights to various products, and there is no question that they can manufacture them efficiently. Yet they cannot successfully market and sell them in America without help from local businesses. Fortunately, we had managed to insert into our agreement with Dimerco a provision for a minimum royalty forcing Dimerco to pay to us even if it never sold the product. It was this obligation that made Dimerco look for an American partner. Chuck Wheatley was able to engage a small product development company called Diablo Research. Diablo had a major customer who was looking for just such a light switch. This customer was Intermatic, Inc., a reputable Illinois company which produced and sold a great variety of high-quality lighting fixtures.

Motion Switch produced by Intermatic

Dimerco made a deal with Diablo, which was later assumed by Intermatic, who contracted Dimerco to produce the wall switch with my motion detector. The Intermatic marketing department dubbed the product the "Motion Switch." Diablo came up with what I think was a clever packaging design, and Intermatic did its best to market it as part of its broad product line of various wall switches. The Motion Switch was sold for $29.95 through such appliance stores as Home Depot, Sears and others. Finally, we began receiving our royalties, yet the first year's sum of about $30,000 went entirely to Henry Goodman to repay his subordinate loan to our partnership. Alex and I received nothing.

The year was now 1985, seven long years since the sensor had first been invented. Infrared heat detectors had finally appeared and were beginning to compete effectively with Intermatic's Motion Switch. As a result, our royalties soon declined rapidly. Nevertheless, every September for the next few years I received a small check from Alex with my portion of the royalties. Around 1990, infrared detectors finally became so popular, and prices for them dropped so drastically, that Intermatic had no choice but to cease production of our Motion Switch.

Even though I managed to make some small amount of money with the Motion Switch, even exceeding my expenses for that invention, I regard the entire enterprise of Bio-Optical Sensors, Ltd. as a lost opportunity. If the switch and the traffic light had been implemented and marketed some ten years earlier, many people might have benefited from these innovations. Apart from being an effective means of energy conservation, these products would have created jobs and generated income not only for me, as the inventor, but for the investors as well. The public, the inventor, the investors—everyone would have been able to benefit from it, if only it had all happened at the right time. Unfortunately, there is no mechanism in this country (or, for that matter, anywhere on this planet) for singling out a useful invention, making it work, and marketing it on a wide scale. Everything is at the mercy of a chance and the perseverance of an inventor.

My next American invention was an apnea detector. The name stands for an electronic instrument which monitors the respiration rate of a sleeping person and sets off an alarm in the event the breathing stops. It had been known for years that an apparently healthy child may often stop breathing for no known reason whatsoever and even die of asphyxiation. This is considered a major cause of a sudden infant crib death syndrome, or SIDS. I had read a great deal about this terrible misfortune, which every year in America alone took the lives of several thousand babies. Usually, a child's life can be saved simply by shaking it if it stops breathing. But you cannot sit by a crib all night long and listen to your baby sleep. It occurred to me that it would be great to have a special detector which could do just that—constantly check that the baby was breathing.

If such a detector were available, it could save many lives. I decided to do something about it, and set off on my own to develop such a machine. It took me many late nights and weekends, but finally I succeeded in devising a miniature radio-transmitter with a special sensor that was sewn into a baby's garment. When this garment was worn by a sleeping infant, the transmitter would send a signal to a nearby radio-receiver in the event the baby stopped breathing. The receiver would then activate an audible alarm to alert the parents to go to the crib and shake the child.

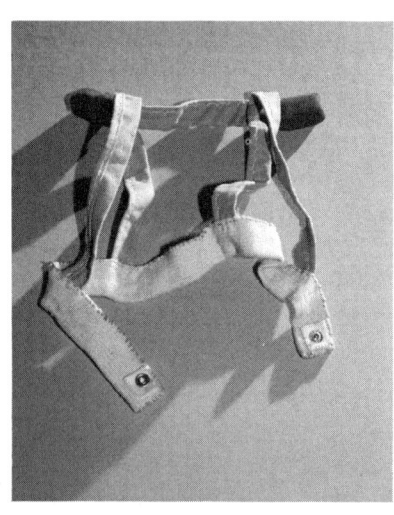

Radio-garment for a baby to monitor respiration

In 1980, my daughter Julia was born and she became my first "guinea pig" for testing the radio-garment. For a prototype, it worked reasonably well, and I began pondering how to put my new invention into practice—that is, how to develop it further into a commercially viable product. Such a task would require a substantial investment and the participation of specialists of many talents. In other words, a new business had to be built from ground zero. To get things going, I felt that I needed a partner with a good understanding of American business practices and an entrepreneurial spirit. Finding such a person has always been the greatest challenge for me, and one at which I have never had particular luck. I was looking for a self-driven entrepreneur on whom I could rely fully. This meant someone who was not only honest, but also a hardworking risk-taker with an understanding of technology and a talent for salesmanship. I was sure that people like that existed, but how did one find them?

We had at the company a young sales manager who had caught my attention. Jim was smart, reasonably well educated, and had a good feeling for high-tech business. In addition, he was a hard-working fellow who was somewhat dissatisfied because his job was not challenging enough. He looked to me like a good choice for a partner in my venture. Frankly, he was the only choice I had. In any event, I talked to him, showed him my radio-garment and suggested we team up. He gladly accepted a 50-50 partnership, and we began developing a strategy.

One day Harry Fein told us that Jim had to go to Moscow for an international exhibition of scientific equipment for biomedical research. The company had rented a booth at the exhibition and Jim was going to present our products there. Besides, we had several customers in the Soviet Academy of Sciences, and it was planned that Jim would meet with them in the hope of obtaining orders for additional instruments. Of course, Harry Fein also would have liked me to go with Jim. I spoke the language and was an expert in the equipment that were to be shown at the exhibition. But this was the period of the Soviet invasion of Afghanistan, and relations between the U.S. and the Kremlin were very tense. It was too dangerous for me to go. If I went and anything happened to me, there would have been little the U.S. government could have done to help me. Thus, my traveling to Moscow was totally out of question, and Jim had to go alone.

At this time, my sister and parents were living in the city of Sverdlovsk, about a thousand miles east of Moscow. My sister had a severe form of diabetes. She needed to take injections of insulin a few times a day. There were two major problems with this: the quality of domestic insulin was very poor, and disposable hypodermic needles and syringes were unavailable. She used an old reusable glass syringe and a couple of hypodermic needles with dull ends which she needed to boil every time she had to give herself an insulin injection. Several times in her letters she asked me if there was any way I could send her a few vials of insulin and a set of disposable syringes. Soviet customs regulations forbade the mailing of medicine or medical equipment to private citizens, so there had been no way I could help her. But Jim's trip to Moscow was an excellent opportunity to hand-deliver a parcel.

Just buying disposable syringes and insulin in America without a prescription was a tremendous enterprise in itself. I will not go into the details of how I was able to accomplish it, but soon I had in my possession ten vials of prolonged action insulin and a box of a hundred disposable syringes. Since our company was planned to ship a large container to Moscow with medical equipment and samples of our microelectrodes, chemicals, and the like, with the Harry Fein's permission I asked Jim to include in the shipment a small box with the insulin and syringes. I told Jim that he would be approached at the exhibition booth by a woman who would take the box from him and pass it along to my sister. I planned to call my aunt who lived in Moscow and ask her to go to the exhibition and pick up the parcel. Even under brutal Soviet conditions, giving medication and syringes to a Soviet citizen was not a crime or even a violation of any regulations, so I believed that neither she nor Jim was at any risk. Jim said that he would gladly do it for me, and I gave him the small box wrapped in a brown paper.

A couple of days later, Jim told me that the container with the equipment and my parcel had been shipped off to Moscow. I then made arrangements to call my aunt in Moscow. This was not an easy task. At the time, making a call to the Soviet Union required placing an order through the AT&T operator exactly one week in advance for a specific time. To get on the list for a call, I had to dial the operator in the first seconds after midnight, otherwise the entire 24 hours of that day's bookings would be taken. With thousands of people trying to do the same thing, the line for the international operator was almost always busy during those first few minutes after midnight. Anyway, after many attempts I placed my order for a call and waited patiently. Shortly after that, Jim drove to Kennedy airport and took a Pan Am flight to Moscow.

A few days after Jim left, Paul Donahue, the company's general manager, walked to my office.

"Does this belong to you?" he asked, handing me a box wrapped in brown paper. "The janitor found it in the wastebasket in Jim's office and thought that it had been thrown out by mistake."

It was my parcel with the insulin and syringes which Jim had told me had already been shipped to Moscow. I was shocked. The man whom I considered my friend and partner had lied to me. Instead of putting the parcel into the container, he had thrown it away, right there in New Haven. I called AT&T and canceled the telephone call to Moscow.

Two weeks later Jim came back. He was very excited about his trip. Moscow was beautiful, the exhibition was successful and he even got a good-size order for the company. Everything was great—except, he said, that no one showed up at his booth to claim my parcel. He told me that he had decided to discard it right there, since there seemed no point in carrying it back to the States. When Jim told me this, I opened the drawer of my desk, took out the parcel and showed it to him.

"Are you talking about these syringes?"

He turned pale and started babbling nonsense, apologizing, telling me that he had been scared the KGB would arrest him, and stuff like that.

"Look, Jim," I said, "if you were scared, I understand that. I would never blame anyone for being afraid to take chances when going behind the Iron Curtain. You should have told me about your fear at the very beginning. That would have been all right. But you lied to me then, and you lied to me today. I am sorry, but I can't trust you anymore."

I told Jim that I no longer wanted to go in business with him and that he should forget about the whole thing. It was not that I wanted to punish him for not taking the package to Moscow. Yes, I was very disappointed and angry, but the loss of trust between us was sufficient evidence that he was the wrong choice anyway. I simply did not feel that I could rely on him in the future.

Jim called me at home that same evening and said that if I did not change my mind, he was going to sue me for breaking our verbal agreement. In years to come, I heard that sort of threat quite a few times from other people of Jim's caliber. A mugger in a dark street attacks you with a knife. A mugger in the office attacks you with a lawyer. I have no patience when someone is trying to frighten me. My usual reaction under such circumstances is swift and decisive, so I just hung up on him. A month or so later, Harry Fein let Jim go for reasons he did not disclose to me. No doubt, Harry had certain problems with him as well.

One day, as I was having lunch in my office, the telephone rang and the receptionist said that there was a gentleman in the lobby who wished to talk to someone from engineering. That was how I first met Floyd Carlson.

Floyd was a tall, white-haired man, well into retirement age, but still quite energetic. He had an enduring desire to do something useful, especially in the area of medical and biological instrumentation. That noble wish, however, had never gotten him very far, for he invariably failed whatever he started. At the outset, nearly all his enterprises looked very promising, but by the end, they always ended up a flop. He began his loser's career in 1948 when he worked as campaign manager for Thomas E. Dewey, who ran for President against Harry Truman. Dewey lost by a hair, and so did Floyd. In the course of his lifetime, Floyd started several business and participated in numerous ventures, all with unhappy endings. His real passion was horses. He spent a good deal of time at racetracks, hanging out with jockeys, trainers, horse owners and veterinarians. There, too, he kept up his losing streak.

Floyd was always looking for opportunities, and it was this that brought him to my doorstep. Initially, I thought that such an energetic and unquestionably honest man might be a good replacement for Jim, but very quickly I realized that I had better limit my involvement with Floyd to the one project he proposed we work on together.

Floyd said that he had an idea for developing a very unusual instrument. He explained that in the horse-breeding business knowing what kind of muscles a foal had was a serious matter. Generally, an animal has two kinds of muscles: the so called fast-twitch and slow-twitch muscles. Apparently, the fast-twitch muscles are capable of fast action for a relatively brief time, for they can store oxygen but cannot replenish it quickly enough. These muscles are very important in horses that race short distances. On the other hand, the slow-twitch muscles cannot move as fast, but they are much more sturdy, thanks to a denser interlacing of capillaries providing a continuous supply of oxygen. In a horse, these two types of muscles function somewhat as in a chicken with dark meat (slow-twitch) and white meat (fast-twitch). Knowing the ratio of these muscles in any particular foal was important in terms of determining the foal's potential and planning its appropriate training. To say nothing of the fact that such information about the contestants could be a valuable tip at the racetrack.

Floyd said that horse breeders and trainers routinely performed biopsies on horses, that is, they took tissue samples from various muscles throughout the animal's body, then froze the samples at cryogenic temperatures and sent them to a special laboratory in England for analysis. He then asked whether I knew of any non-invasive test (that is, a test that did not involve penetrating the skin) to replace the painful, costly and not terribly efficient muscle biopsy. Ideally, such a test would give a percentage of fast-twitch versus slow-twitch muscles. At first, I said that I had never heard of anything capable of doing what he needed, but after giving it some thought, I came up with an idea which seemed worth trying.

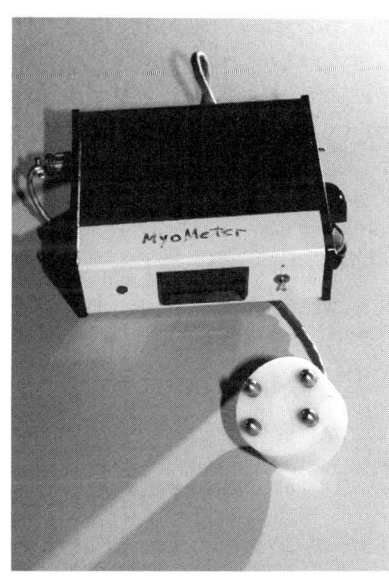

"Myometer" for measuring properties of muscles

I did some reading, sketched out a few schematics, and then called Floyd. I told him that I could build a test prototype to see whether my idea was feasible. Floyd agreed to pay for the parts and I got down to work. Within a month I had built an electronic box with a two-foot cable. At the end of the cable was a probe resembling a stethoscope which had a number of the electrical contacts. The box had a gauge with a scale graduated by percentage of fast-twitch muscle content. I called the apparatus a "Myometer" which meant a device for measuring muscle fiber. To see what kind of muscle was at a particular spot on the body, I pressed the probe against the skin for a couple of seconds, pushed the "Measure" button and checked the gauge. It was quite simple. I started out by testing the instrument on myself, and it showed that I had different kinds of muscles all over my body. But what kind and how different? Measuring myself was not sufficient for testing the prototype—I was not a horse, after all. The instrument needed calibration and a comparison with actual biopsy tests.

Floyd suggested that we go to Belmont Racetrack near New York City, where he had several friends among the horse owners and trainers who would surely allow us to do some testing on real horses. We drove to the racetrack and Floyd arranged for me to take measurements from five horses whose biopsy results were already known. I made about a hundred measurements and was pleased to see that the gauge indeed showed very distinct readings for fast-twitch and slow-twitch muscles.

I gave the instrument to Floyd and taught him how to collect more data. He did so quite successfully over the course of the next few weeks. The apparatus worked, but to proceed further with it and make it into a practical device for the market required more thorough testing on a larger number of animals. Its circuit had to be refined to simplify the operation, the probe shape had to be improved, and so on. It was clear that a lot of work was needed before we had a commercially viable product. I was ready to do my part and expected that Floyd would take care of the business side, that is, raising money, developing a business plan, figuring out a sales strategy, distribution, and so on. Unfortunately, he was a great talker but a poor doer. Our collaboration never went beyond mere talk and daydreaming.

While working on the instrument for measuring horse muscles, I began to think about other body tissues. What about malignancy? If I could detect what kind of muscles lay underneath the skin, could I also detect the presence of cancer cells? The result of these musings was a design for a personal breast cancer detector. The prototype I built in my home laboratory resembled a cylindrical flashlight. Its consisted of a sensing probe with sixty-four miniature contacts which, when pressed against the breast, sent electric currents of extremely small magnitude through the skin. These currents passed differently through benign and malignant tissues, and the built-in microprocessor distinguished the one from another. In the event of a suspicious growth, a red light came on. If nothing abnormal was detected, a green light came on. The idea was that a woman could use the detector on herself at home in privacy, and if she noticed a red light at any point while examining her breast, she would then go to her doctor for a more thorough examination.

Breast malignancy detector

I built the prototype and naively hoped that my new invention would be welcomed by both doctors and consumers. I started actively raising money for further work on the instrument. This invention was a serious matter. Unlike the Miometer, it would require extremely elaborate clinical testing and engineering refinements. My first prototype was good enough only for the initial feasibility study. It was obvious that the breast cancer detector could be further developed only in cooperation with a respectable research hospital and would require fairly substantial funding.

Many months of an unsuccessful search for a possible investor led me to the conclusion that the development of a medical device whose use, or rather misuse, could potentially draw a fine line between life and death, was a hopeless enterprise. Practically everyone to whom I showed the prototype said that, if proved effective, the invention could be an extremely successful consumer health care product. But, at the same time, everyone recognized that demonstrating its validity would require years of work and millions of dollars, not to mention a difficult struggle with the FDA to obtain approval of the detector for consumer use. It is little surprise that no one was willing to put up the money for its development. It was quite a risky investment, indeed. I had no choice but to put this invention to rest alongside so many others.

During the early eighties, I developed several medical inventions which never grew beyond the initial prototypes. Now, ten years later, I still believe that every one of them had good potential and, if properly funded and managed, could have led to a successful business venture. Here, I will briefly describe a few of them.

One invention was something I called a "UV-detector." It was a personal health care device the size and shape of a wristwatch. A built-in semiconductor sensor monitored the intensity of ultraviolet rays in the ambient sunlight and triggered an alarm if the cumulative level of UV surpassed a preset safety threshold. I believed such a detector could be a useful device to assist in the prevention of skin cancer. Unfortunately, as with my breast cancer detector, I was unable to find a sponsor for the invention and laid it to rest in my suitcase of unrealized dreams. Apparently, the idea of monitoring UV intensity has since caught on, because a few years later I saw a very similar detector in a catalogue of consumer products. Someone else had had the same idea and was more successful then I in making the product fly.

Another idea was suggested to me by my friend Professor Simon Gelman, who today is chairman of the anesthesiology department at Harvard Medical School. In the mid eighties he drew my attention to a certain problem in anesthesiology—the frequent necessity to monitor the depth of anesthesia of a patient undergoing a surgical procedure. In many respects, anesthesia was more of an art than a science, and the possibility of determining quantitatively the physiological response to the administered drugs could be a very attractive feature.

Simon and I decided to develop a noninvasive anesthesia monitor which would resemble an electrocardiograph with small electrodes to be placed on the patient's fingers. The monitor would analyze the shapes of the pulsating blood flow by measuring the electrical impedance of body tissues. A microprocessor would calculate the specific parameters of these shapes and relate them to so called peripheral vascular resistance, which, in turn, was related to the depth of anesthesia.

For that project, I already had the cooperation of a fine doctor and the potential support of a respectable medical school. Nevertheless, in the end it failed to work out as well. By the time I had built a working model of the anesthesia monitor and tested it on myself, another of my inventions (the instant thermometer) was progressing quite successfully and I ended up having to spend most of my time working on that project and raising money for its commercialization.

I wrote several proposals for grants to fund development of the anesthesia monitor, but they were turned down. Obviously, Dr. Gelman was disappointed that the project went nowhere, but there was nothing I could do about it. I had only one life and only 24 hours in a day. The investors who had entrusted me with their money quite rightfully demanded that I focus my efforts on a project which carried less risk of failure and greater chances for financial success. The anesthesia monitor, or, for that matter, any life-supporting or diagnostic device, did not fit that category.

Probably, the only feasible way to introduce such instruments into practice, would be to develop them through a university laboratory in cooperation with a research hospital, while the best source of funding for these products would be government grants. After all, it is the responsibility of a society to see that medical instrumentation is progressing and benefits everyone.

Any investor loves an invention that would entails the production of "razor blades," that is, disposable devices which can be used only once and then must be discarded. Every time you need such a device, you have to get the new one, like a razor blade. A steady disposable business is a money-making machine and an investor's dream. Another of my inventions (one which I actually made in Russia long before coming to America) was a disposable medical device. Anyone who has spent some time in a hospital or seen how paramedics work, knows that the most often monitored vital sign is an electrocardiogram, or EKG for short. It is an essential tool for monitoring the activity of the cardiac muscle. The instrument for recording an EKG is called an electrocardiograph and was invented in 1903 by the Dutchman Willem Einthoven. To pick up the minute EKG signals from the surface of a patient's body, several electrical contacts, called electrodes, are attached to the skin and connected to the electrocardiograph by a cable. This cable often becomes a big pain in the neck for medical personnel. Whenever the patient has to be moved, the cable must be disconnected, thus interrupting the ability to monitor. This is why, outside operating rooms, patients are connected to bedside monitors only in intensive and critical care units.

While working with bio-telemetry systems, I came up with the idea of making a wireless EKG electrode. This electrode would be almost the same size as that of a conventional electrode, which is a small conductive patch with adhesive to stick it to the patient's chest. Inside the patch I wanted to implant a microchip with a tiny radio transmitter. The transmitter would send the EKG signal the short distance to a bedside monitor. Such an electrode could stay on a body for at least a week, transmitting the EKG continuously, day and night. The patient could sit, stand, walk about, and even take a shower without worrying about the electrode, and his EKG would still be monitored and analyzed.

I built the EKG-transmitting electrode using a small lithium battery which had sufficient power to broadcast heartbeats continuously for several months. Whenever I visited potential investors, I attached the electrode to my chest under a dress shirt. During my presentation, after telling them about the invention, I used to say, "Gentlemen, would you like to hear my heart?" I would then take from my pocket a small AM-receiver I had bought at Radio Shack for $9.99 and turn it on. The electrode transmitter was tuned to the standard AM broadcasting range, so that at each beat of my heart, the radio-receiver produced a distinctive sound, something like "bleeoo-blip, bleeoo-blip." This usually made a great impression on people with deep pockets. They always liked something they could understand and that had a certain novelty-like feature.

10

Fight for Light

> O Lord,
> Thou givest us everything,
> At the price of an effort.
>
> *-Leonardo da Vinci*

I remember from my childhood the day Stalin died—March 5, 1953—and my mother sobbing uncontrollably, "How we are going to live? Who will take care of us?" The god who had been expected to live for eternity had suddenly died, and mortals thought that the end of the world had come.

Along with millions of ordinary Soviet citizens, my parents were brainwashed and could not imagine any way of life other than as slaves of the state. An insect does not suffer from its being an insect, for it does not know what it is to be a bird, or a fox, or a human. Likewise, those who had spent most of their lives under one of the cruelest regimes history has ever seen, were glad merely to be alive and have bread on the table. They knew nothing better and could not even contemplate a different life for themselves, either there or anywhere else on earth.

When I decided to leave my homeland, it was a moral shock for my parents, especially for my father. They felt that I was venturing off into a deadly abyss from which there was no return. My parents believed, quite rightly, that we might never see each other again. The law of the land dictated that I could not go abroad without written permission from my parents, and squeezing a signature from my father was no small matter.

Once we crossed the border, I made a new habit for myself. I began writing letters to my parents on a regular basis—one letter a week. Over the course of nearly nine years, until they themselves managed to come to America, I sent them 395 letters, the carbon copies of which I still keep.

Of course, the KGB opened every single piece of correspondence going in either direction. I developed a method of checking to see whether a letter I received from my parents had indeed been opened. Of the 357 letters I got from my mother, only two were not opened. In writing to each other, we had to be careful what we said and how we said it. We were our own best censors—probably far more vigilant than the paid KGB employees who had to read hundreds of similar letters each day. Even so, in every one of my letters I told my parents something about America and life in the free world. It was as if I were creating a mosaic, dot-by-dot and piece-by-piece, until gradually I painted the full picture. After a few years of my epistolary teaching, their cleanly washed brains grew humanly murky with a far broader knowledge and understanding. Nevertheless, my parents continued to live the lives they had always known there, working and then retiring. But everything suddenly changed when my sister lost her eyesight.

Because of her diabetes, Marina had had problems with her eyes for a long time. Every year she had to go to the Helmholtz Eye Institute in Moscow to undergo retina reattachment. Then one day a visiting doctor from the Soviet republic of Georgia suggested that she try an experimental treatment which, he told her, would permanently reattach the retina and correct her eyesight forever. Believing him (because she wanted to believe him), she flew to Tbilisi, the capital of Georgia, and the doctor operated on her eyes. When she returned home to Sverdlovsk and the bandages were removed, her vision was clear and she resumed her work as a physician at a local hospital. But a few days later, a massive hemorrhage started in her eyes. For a time she could see the world as if through a red glass, but before long the blood inside her eyes coagulated and darkness fell.

My mother took her again to the Moscow eye institute. Marina spent a long time there, but with no improvement. One of the doctors told her that they had neither the equipment nor skills to repair her eyesight. He mentioned, however, that an American eye surgeon from Johns Hopkins University in Baltimore, who had recently spent some time at the Moscow Eye Institute, had said that similar cases of hemorrhages in the eye were quite treatable in America.

I called the university and found the doctor who visited Moscow. He told me that his trip had left him with little respect for Soviet eye medicine, and that it was no surprise to him that the surgery had resulted in blindness. When I asked him what, if anything, could be done, he replied, "First, let me see the patient, and then I'll tell you what can be done." He was right, of course—he had to see the patient. But that was the biggest problem of all: how were we to get my sister to America?

Marina and my mother went to the local authorities in the Urals to seek permission to travel to America for medical treatment, but they were met with ironic smiles. The Soviet war in Afghanistan was at its height, relations with the United States were at their lowest level ever, and the détente was dead. The Iron Curtain was once again impassable, and all the efforts of my mother and sister to get permission to travel to America resulted in nothing.

Meanwhile, I tried to do whatever I could from this side. I contacted numerous congressmen and senators, but with little success. Then one evening, I saw a TV show where an old but still lively man was talking about his unique connections with the Russians. His name was Arman Hammer, and he claimed that in spite of all the wars, hot or cold, he had remained a friend through sixty years of all the Soviet rulers from Lenin to Brezhnev. He said, "Even today, when there is no dialog between our two countries on a government level, I am the only American who is always welcomed in Kremlin and I am the only voice from this country to whom Brezhnev listens."

Upon hearing this, I thought, why not ask Dr. Hammer for help? I prepared a very carefully composed letter which I marked "Personal and Confidential," and mailed it to Occidental Petroleum, where Hammer served as chairman. In the letter, I described the situation and asked him to use his influence to obtain a travel permit for my sister and mother, mentioning, of course, that I was not asking him for any money and would pay all the expenses myself. A long time passed after I sent the letter to Occidental Petroleum. I waited and waited and waited. Finally, after some four months, I received a response from Arman Hammer in which the old braggart wrote me that "...due to the present tension between the two countries, nothing I can do to help your sister."

My other attempt to use influential people was through my friend Floyd Carlson. Floyd knew the chairman of Pepsi Cola, whose name was Don Kendall. At that time, Pepsi was the largest business partner of the Soviets, and Kendall, like Hammer, was a frequent visitor to the Kremlin. Floyd wrote a letter to Kendall asking him for assistance. Unlike Hammer, the chairman of Pepsi Cola responded quickly, yet his letter was a perfect example of either infantile naiveté or sheer stupidity. Literally, in the letter he said, "Floyd, why on earth do I have to get involved in such a small matter? Everything can be resolved much easier without my bothering Mr. Brezhnev. I suggest that your Russian friend should contact the Soviet embassy in Washington and ask the ambassador, Mr. Dobrynin, to issue a travel permit for his sick sister." Surely the influential chairman should have know that for an immigrant like me to ask the Soviet ambassador for help would be as effective as asking the IRS not to collect taxes.

When our efforts on both sides had failed, my parents and sister decided that the only option left to them was to emigrate from the Soviet Union for good. We hoped that there might still exist some chance for her to see light. However, in 1982 emigration became extremely difficult. Only a handful of people were allowed to leave the country, and the city where my parents lived was completely closed to foreign visitors. We even felt lucky that our correspondence was not interrupted. After considering all cons and pros, we decided that they would have to move to another city, just as my parents-in-law had done two years earlier. Preferably, they should go to one of the Soviet republics, were the central control was weaker and people were more sympathetic.

Changing one's place of residence in the Soviet Union was almost as difficult as changing one's country. Only if one could arrange an exchange of living quarters with an equal number of people from another city was it allowed. My mother began the truly heroic enterprise of finding someone at the periphery of the gigantic empire who wanted to move to the Urals and into the apartment where my parents lived with my sister, her husband, and their child—along with my eighty-five-year-old grandmother.

To make a long story short, after an extremely painful and frustrating search, my parents were able to arrange an exchange and moved to a filthy apartment at the outskirts of the Georgian capital of Tbilisi. After the relocation, they resumed their efforts to obtain permission to leave the country. Though still very difficult, humiliating, and tiring, such efforts in Tbilisi were not totally hopeless.

Here on this side of the Iron Curtain, I had to arrange for them an invitation from me as though I were a citizen of Israel, for only invitations from the Jewish state were honored by the Soviet authorities. It was part of the anti-Semitic propaganda to claim that only the Jews desired to leave with the "workers' paradise." Since I was not a citizen of Israel and indeed had never been to that country, there was no way for me to obtain the official papers. In the past, Israel, in order to save people from oppression, had issued numerous invitations to Soviet Jews from fabricated relatives in Israel. I myself had made use of just such an invitation when I applied to leave the country. But later on, when the Israeli government realized that most of the Jews who emigrated from the Soviet Union settled in other countries, primarily the United States, this practice had been stopped. Obtaining such an invitation became, therefore, a real challenge for me.

My brother-in-law Misha Tyshkov and his wife Masha, who at the time were already living in New York, told me that they had heard of some kind of underground organization which specialized in forging documents. Such a criminal group did indeed exist. I did not know, of course, that it forged not only documents of various sorts, but dollar bills as well. It was not until a year or so later, when the U.S. Secret Service raided their printing shops and made numerous arrests, that I learned from the newspapers how advanced the group was.

I asked Misha to sniff around and to find out how I could go about contacting those people. I wanted to ask them to fabricate an "official" Israeli invitation for me. My conscience did not bother me at all, for I had no intention of breaking any American law. I wished to make a fake document of another country (Israel) to be used in a third country (the Soviet Union) for a quite noble cause.

A couple of weeks later Misha called to tell me that he had talked to a Brooklyn shoemaker who claimed to have access to people who could do it for just $1,000. The money had to be paid in advance—in cash, obviously. I agreed, got the money, and drove to New York. Misha took me to a tiny shop where a middle-aged man was nailing heels to middle-aged boots. I waited until no customers were around and asked the shoemaker if he could arrange for an invitation to be made for me.

"I know exactly what you want. Did you bring the money?" he asked, keeping the shoe nails between his lips. I handed him an envelope with the cash and a piece of paper which listed the names, address, and birth dates of my relatives. Without looking at it, he quickly put the envelope into his back pocket and said, "Okey-dokey, call me in a month. I'll tell you if it's ready."

I called him four weeks later, and he said that everything was ready and I could stop by to his shop any time to pick it up. To save me a trip to New York, I called Misha and asked him to go to the shoe shop and get the document from the shoemaker. Misha retrieved it and mailed the paper to me.

It looked perfect. A genuine document with all the watermarks and official seals. There were even actual duty stamps attached above the signatures in Hebrew letters. It was a great job, and I could not even tell whether the paper was forged or those people had used actual Israeli forms and the real stamps. But suddenly, I noticed that the birth date of my sister's husband Yury was wrong. Instead of 1940, the invitation had 1941. This error essentially invalidated the entire document, for the Soviet authorities would surely use the discrepancy as a reason to turn down the application. The document had to be corrected.

I called Misha and asked him to go back to the shoemaker and ask him to redo the document. Misha called me back the next day and said that the correction would cost another $1,000. I decided to go to Brooklyn myself and talk to the man. My wife warned me that it might be dangerous and asked me to hold my temper.

When I walked into the shop, the shoemaker was alone.

"Look," I said, showing him the document. "Whoever did the job, made a mistake. The date of birth must be changed. I paid the money and expect to have a document with no errors. Will you remake it, please?"

"Didn't I tell your brother-in-law that if you want another document, you must bring another thousand bucks?"

"But it was *your* mistake, not mine," I began to protest.

The shoemaker's face turned red, "Listen, you asshole! I am just a delivery man. Got it? I don't do these things myself. My boss told me—either bring another grand, or nothing. Now, beat it!" With that, he opened a drawer in his bench, took out a revolver and pressed the muzzle to my forehead, right between my eyes.

"If you come here again and nag me, I swear to God, I'll put a slug in your head."

The gun was a very convincing argument, one to which I had no response. I walked out of the shop and drove back to Connecticut. The situation was quite desperate. I saw no other way but to get another thousand dollars, which I did not have. I decided to give it another try and call the shoemaker on the phone. I had little hope, but at least he could not shoot me over the phone line. I dialed the number and had barely managed a "hello" when he started shouting and threatening me all over again.

"I'm telling you for the last time, if you show your face in my shop, I'll put a bullet right between your eyes. Mark my words!"

I responded, "Yes, I'm marking your words all right. I've even recorded this conversation on a tape recorder, just in case."

With those words I hung up. Of course, I was bluffing. I had not recorded a thing and was merely telling him that in desperation, just to get back at him.

About ten minutes later my telephone rang. It was the shoemaker. This time, his voice was soft as a baby's bottom and sugary as Nutrasweet.

"Let's talk like two gentlemen," said the gentleman-shoemaker. "You were angry, I was angry. Please understand. I'm a middleman, I just take it here and give it there. What can I do? It'll cost more money. The best I can do is three hundred. That's not much. Well, what do you say?"

I said okay, went to New York, paid him the three hundred, and in another four weeks had my corrected invitation. Delivering it then to Tbilisi was a whole other story. Naturally, I could not trust the regular mail, for the invitation would almost certainly be intercepted and never reach my parents. Instead, I found someone who was traveling to Russia and persuaded him to take the document with him. My mother received it and filed all the necessary papers with the authorities. I was very nervous about how it would go through and just prayed that everything would work smoothly.

All these arrangements took a long time. We were afraid that the more time that passed the less chance Marina had to have at least a portion of her eyesight restored. Already it had been nearly four years since she had seen light for the last time.

In December 1985, exactly eight years after my arrival in America, my parents, grandmother and Marina's family received permission from the Georgian officials to emigrate to Israel. Shortly thereafter, they left the country and in January arrived in Vienna.

My now ninety-one-year-old grandmother was in a very bad shape. The night before they were to fly out of Russia, she lost consciousness, and everyone thought that she was not going to make it. My mother injected her with some very powerful medication, while Marina stuck needles all over the old woman's body (my sister had caught from me a love for Chinese folk medicine and had become a skillful acupuncturist, which was all the more amazing, given that she was blind). These various remedies helped and my grandmother was able to travel to Vienna.

I love surprises and wanted to arrange a surprise for my family, so before they left the country, I told them on the phone to be at the steps of the Viennese Opera at 11 A.M. on February 25. I said that someone would be there to bring them a present from me.

All six of my relatives made it safely to Vienna, where for some time they lived in a small hotel for refugees. On February 25, they went to the Opera steps to meet the messenger (my grandmother was too weak and stayed behind at the hotel). Naturally, it was I who came to meet them. After almost nine years, the reunion was very emotional and touching.

We took the streetcar to their hotel. My grandmother did not recognize me at first, but when I came closer, she began to sob. I comforted her, and then she looked me up and down with an appraising eye. I was well dressed. My jacket had fancy leather patches on the elbows, which made a totally wrong impression on her.

"Oh, dear, you are not doing well in that America. Your clothes have patches..." And she began weeping again.

We spent two days together, then I took the train to Rome, followed a few days later by my family. They had to spend several months in Italy before an American immigrant visa would be granted. For four days we walked together all over the ancient city, visiting the Vatican and many museums and parks, and at the end of the four days I flew back to the States.

Since I was an American citizen, my relatives could come to America only if I guaranteed their welfare and pledged to support them fully, so that they would not be a burden to the taxpayers. I drove to Hartford, where the Immigration and Naturalization Service was located. There, I had to file an affidavit of support for six people. The clerk who took my application said that such an affidavit could only be accepted from an immediate relative. I said that the people were my parents and my sister and my sister's husband and son.

"That's close enough," responded the clerk, "but you must show us proof that these people are indeed your mother and father. I need to see your birth certificate with their names."

I explained her that when I left the Soviet Union I had not been allowed to take any documents with me, and that I therefore had no such papers available.

"In that case, we'll need two witnesses."

"Oh, that's easy," I said. "I know of several people in the United States who can sign statements that they have known these people as my parents for many years."

"No, that's no good. We can accept only witnesses who were physically present at the moment of your birth."

"What do you mean at the *moment*? Are you telling me that the only witnesses you can consider are the midwives?"

The clerk shrugged her shoulders, "Basically yes. It's up to you, though. I can take your petition, but without acceptable proof, your chances of having it granted are next to nothing."

What could I do? I signed the papers and went home.

I did everything I could to try to get the bureaucratic wheels and gears to turn faster. I called the State Department, wrote letters to congressmen and senators. I begged them to expedite the issuance of an American visa for at least my sister. She was out of Russia, but still too far from the eye doctors I wanted her to visit. Everyone responded to me with compassion and pledged to do something. Congressman Bruce Morrison promised to use all his influence and keep me posted on the results of his efforts. Six months later I received a letter from him in which he informed me that the visa would be granted shortly. His attempt was a good one but a bit late, as at that time my relatives were already living in America.

Meanwhile, my folks settled in Ladispoli, a small resort town on the Tyrhenian Sea. It was winter and the rents there were cheap. They found a nice apartment and tried to enjoy their Roman holiday. Italy is beautiful at any season. Rome was not far away and they traveled there often.

Every day, my poor grandmother felt worse and worse, and one night my mother had to call for an ambulance. The old woman was taken to a local hospital where she died, just one month shy of her ninety-second birthday, never having fulfilled her dream of coming to America. Her life had been a difficult one, even tragic. During the Stalin's artificially created famine, she had nearly died of starvation. Her husband, my grandfather, spent many years in the Gulag, and she fought a perpetual battle to bring up her children. In 1942 her son was killed by the Nazis. All her life she worked hard, helping her hands with her brain. Now, finally, she rests in peace in the Eternal City.

Three months later I received a letter from the Immigration Service. The letter informed me that for the purposes of the immigration law my parents could not be regarded as my parents, and that my affidavit of support was declined.

I had expected that kind of response, but still could not believe that American bureaucracy could be so absurd. The rejection simply meant that my parents would have to retain their refugee status and arrive in this country through the full support of the U.S. government, as if I did not exist. I was ready to pay, but Uncle Sam did not want my money. I made a copy of the letter and sent it to the American consulate at Rome, which was processing the visa papers for my parents. It all took some more time because of the delay due to the illness and death of my grandmother, but finally, on August 26, I went to Kennedy airport to meet my awfully tired but immeasurably happy relatives.

Soon after their arrival, I took Marina to a New York eye clinic, then to an eye surgeon at the Yale New Haven Hospital. The doctors all told us that too much time had passed, and that the chances of restoring her vision were extremely small. The Yale doctor suggested doing an exploratory operation. He said that there was no more than a 10% chance that he could salvage one of her eyes. Stoically, Marina went ahead with it. The surgeon opened one eye and found that the connective tissue had grown through the entire eyeball and that nothing can be done.

Marina learned to live in darkness. She is now in Connecticut with my parents and her husband Yury and son Eugene. She received a license as an acupuncturist and masseuse and opened her own therapeutic massage business. She is a workaholic. I consider myself a hopeless optimist, but in that respect Marina beats anyone, including me. She is exceptionally energetic and joyful, with an incredible zest and passion for life. She has numerous friends and teaches us all a lesson in stoicism and buoyancy.

11

High Pressure

> The difficult we do immediately,
> the impossible takes a little longer
> *Fritdjof Nansen*

A couple of months after I split with Jim, I received a phone call. It was Al Willis, the director of research and development at Timex, the watch company. He invited me to come over for a job interview. He said that I had been recommended to him by the Case Western Reserve University in Cleveland. I told Mr. Willis that I was not looking for a job and was not interested in making watches, but he insisted that we meet. I decided to go, if only to hear what he had to say.

Timex was located about 25 miles from my home in the picturesque town of Middlebury, Connecticut. I put prototypes of several of my inventions into my briefcase and made the drive to Timex. It took me about an hour to get there through the charming and winding country roads. At Timex, I was interviewed by some four or five people, of whom the highest ranked was the vice president of R&D. They told me that Timex was about to open a new division for the development, production, and marketing of high-tech home health care products. They were looking for a senior biomedical engineer with a strong background in the monitoring of vital signs. This person was to take care of the research and development of some of the new products.

The first project Timex wanted to develop was a crib alarm detector. Upon hearing this, I opened my briefcase and took out my radio-garment. It made a dramatic impression on everybody there. My interviewers could not believe that I already had a working model of this product that Timex had only just begun talking about. I demonstrated how the prototype worked and my hosts liked what they saw. Before the meeting was even over, the director of R&D made me an offer, which I declined. I told him that I was happy with my present employer and saw no reason to switch jobs. Al Willis then asked what it would take to change my mind. To end this discussion, I decided to ask for something I was sure was totally unreasonable.

"Double my present salary," I said, "and I'll think about it."

The immediate answer was yes. Caught off guard, I asked for the offer in writing and said that I still needed a few days to think it over. When I got home and told my wife about the offer, she said it would be foolish not to consider it seriously. Of course, I was excited about the prospect of being involved in a new consumer medical business as well as the chance of improving our standard of living, but I hated the idea of commuting so far and did not want to abandon just then the development of a new piece of equipment which was under way at my present place of work. I knew that my leaving would be a serious blow to both Harry Fein and his small company.

I realized that joining Timex and developing the crib alarm for them would definitely mean giving up any hope of striking out on my own, at least with that particular invention. My major concern about going into business by myself was that I had no partner, no business experience, no money, and no contacts in the business world. Besides, knowing that an organization as large as Timex was planning to launch an identical project, meant that I might face tremendous competition and my chances of success would be very slim. In this respect, joining Timex seemed to make sense. Another attractive aspect of being at a large company and actively participating in a project which I had already started on my own, was that it could propel me to a new, higher professional level.

On the other hand, I had no desire to terminate my present project, which was well under way, nor did I want to cause any harm to my then employer. In short, there were plenty of pros and cons to the offer. Unconsciously, however, I wanted a change and was searching for good reasons to convince myself that I should accept the offer.

Now, many years later, I know that an entrepreneur should never worry too much about competing with a large company, at least in the research and development phase. A large body has great inertia—not much of a benefit in a race, if it is a race run along the twisting road of invention and product development. If it was a straightforward engineering process, manufacturing, and marketing, that was a different ball game.

At that time, I had already been working for Harry Fein for two years and had asked him several times to give me a raise. He had hired me cheaply, comparing with what he would have had to pay an engineer with an American background. When I got the job, I was almost a newcomer, with no experience in industry, and had agreed to work for a lot less. However, when I saw how quickly and profitably my designs were implemented by his company, and how nice a profit Harry was making, I asked him to give me a raise. He said, "No, I won't give you a raise, but at the end of the year you'll receive a pretty good bonus." I said okay, and went on working as before.

When I got Timex's offer and showed it to Harry, I said, "Harry, match this offer and I will remain with you."

"I wish I could, but I can't," he said. "Look," he continued, "if you leave me now, before the project is finished, you'll kill me. I've invested so much money in it and your leaving will be a disaster. Haven't I been good to you these past years? Please, stay with me."

"Okay, Harry," I said, "I don't want you to be in trouble, but, on the other hand, I can't pass up this opportunity. This is what I am going to do. I'll call Timex and tell them that I am accepting their offer, but only after three months from now. During these three months, I will finish your project, so that you'll be able to take it straight to the production floor, with no further need for my personal participation. Of course, during these three months I will be losing a lot of money in my paychecks, but you promised to give me a bonus, didn't you?"

"Of course," said Harry, smiling happily. "Thank you. You are a good friend."

The next day I called Timex and told the director of R&D that I accepted the offer, but would not be able to join them for another three months, as I had some obligations with my present employer. Al Willis was disappointed and tried to persuade me to shorten that time, but finally agreed to wait.

The following months were probably the most intense in my professional life. I worked day and night, and within about nine weeks the entire project was finished, the prototypes were fully tested, and all the technical documentation was in a ready-to-go condition. Harry was happy, shook my hand and thanked me heartily. Then I said, "Okay Harry, I did what I promised. I am leaving. Now, where is my bonus?"

"What bonus?" he asked. "It's not the end of the year yet. Stay till Christmas and I will give you your bonus."

Once again I had learned that no good deed goes unpunished.

I left his office with a burning desire for revenge. In completing the project, I had transferred every last detail of it to my technician. Ray was the only man at the company who would be able to carry on the project after my departure. So before walking out the door, I made him an offer.

"Ray, come with me to Timex," I said. "I'll get you an interesting job and a better salary." He agreed gladly and we left together. Harry was furious, but I believe he deserved it.

High Pressure

> If you have a job without aggravations,
> you don't have a job.
>
> -*Malcolm Forbes*

Timex, or, more accurately, the division of the company called Timex Medical Products Inc., was the first large American corporation for which I worked. During my first year there, I enjoyed my work very much. The job was challenging, the technical support was strong, but most important of all, I had around me excellent engineers from whom I learned a great deal. Timex was owned by Fred Olson, a Norwegian man with a strong entrepreneurial spirit. When he noticed that competition from Asian companies was reducing his market share in the watch business, he decided to diversify. To do this, he formed several corporations under the Timex umbrella and directed them to develop consumer products in different areas. One was involved in the production of a 3-D photographic camera, another in that of a personal computer (a very innovative product in 1981), and a third in self-care high-tech devices. In reality, all three daughter-companies were situated in an enormous room with partitions, and the division was rather organizational than physical. In the same room there was also a Timex department for the development of gyroscopes—directional sensors for missiles and aircrafts.

Since Timex had some expertise in the photographic camera business (at the time, it produced all the cameras for Polaroid), Fred decided to obtain a license from two French inventors to produce and sell a 3-D Nimslo camera. It was a terrific invention. The camera had four lenses and required the use of three frames of film to make one picture. Once the pictures were taken, the film had to be sent to a special lab for processing. The photographs looked like the popular 3-D postcards with a corrugated plastic surface, only you could make them yourself. Timex priced this unique camera quite modestly—about $200—and made various preparations to launch the product.

I was curious about the Nimslo and wanted to get a feel for how it worked. After all, I had over twenty-five years of experience in photography, with both still and motion picture cameras, so I believed I could pass some judgment on that product. I asked one of the designers to lend me the camera for a weekend. The designer explained to me that taking 3-D pictures required careful planning and composition. I bought a 36-frame roll of color film (to make a dozen 3-D pictures) and, on a sunny day in my backyard, took twelve shots. Supposedly, I did everything right, exactly as the designer had instructed me. I placed the subjects (my children) at different distances from the camera, made sure that the lights and shadows provided sufficient contrast over the depth, and made other equally important arrangements which one generally ignores when using conventional cameras.

After finishing the roll, I sent it to Atlanta, Georgia, where the processing laboratory was located. I awaited the return of my first 3-D photographs with great anticipation. But when the packet with the prints arrived, I was quite disappointed. Only three or four pictures had any really noticeable depth effect. The others were not impressive at all and generally of a lesser quality than conventional color prints. This meant a lot to me. Given my experience in photography and detailed instructions from someone who had actually worked on the mechanical design of the camera, my output was very poor. I should have expected at least eight good pictures, not four.

I came to the conclusion that the Nimslo camera was a wrong choice for the consumer market. No matter how good the design, it was unrealistic to expect the average consumer to be satisfied with so poor an output at such a relatively high price per shot. Surprisingly, the designer shared my opinion, but said that the marketing people did not want to listen to such arguments.

Nevertheless, I decided to give it a shot. I went to the vice president of marketing, showed him my pictures and said that I predicted a great disappointment in the marketplace, if not a disaster. To that, his response was: "Fred [Olson] wants it, the company wants it, I want it. I hear what you are saying, but, frankly, you come from another world. You don't know the American consumers. You'd better stick to your engineering business—make us a good crib alarm and leave the marketing decisions to the professionals."

I decided he might be right, that maybe I really could not see the whole picture. So I went back to my lab and forgot about Nimslo.

Shortly thereafter Timex put the camera on the market. The initial response was favorable, but before too long, the return rate of the cameras by consumers exceed sales. One year later, Timex ceased production of the 3-D cameras and sold all the rights to a Japanese company. The Japanese wisely redesigned Nimslo for professional use, that is, to produce special 3-D effects in photographic studios. Timex lost a great deal of money, and I did not feel like walking around with a gloating smile and saying, "I told you so!" The bad luck with the camera was one in a series of failures with truly innovative products which Timex sought to produce and sell.

Another new product was a personal computer. Timex had bought the rights to it from a British inventor and entrepreneur Clive Sinclair, who launched the first low-cost ZX80 computer in 1979 and introduced a color version called Spectrum in 1981 (I still have this machine in my collection). For his achievements, Sinclair was knighted and is now known as Sir Clive. Indeed, the British know how to recognize their great innovators.

In its worldwide search for high-tech consumer products, Timex decided to gamble on Sinclair's personal computers—not a bad idea. The company hired several hardware and software engineers, put them in the same large room with everybody else and directed them to modify the machine for the U.S. market. And again, as with the camera, the product experienced an initial leap followed by great fiasco. (In this instance, however, I had no personal opinion and was as sorry as everyone else when Timex lost out to Apple, its major rival).

Besides myself, there were several other engineers in the medical division who were working on less sophisticated products, namely, an oral electronic thermometer and a bathroom scale. They were all either mechanical or electrical engineers, while I was the only biomedical engineer. As had been planned, my task was to develop a crib alarm.

After careful testing and analysis of my initial idea of using a tiny radio-transmitter sewn into a baby's garment, I abandoned this in favor of a more attractive idea. I decided to devise a totally noncontact detector. I did not want to attach anything to a sleeping baby. To measure breathing, the sensor had to be hidden inside the crib, not on the baby. One of the engineers, Jack Schwarzchild, told me that a year earlier he had come across an interesting polymer film produced by a Japanese company, Kureha. The film possessed piezoelectric properties, that is, it could generate an electrical signal in response to mechanical stress. This sounded like a very interesting property and one which I could put to good use. Jack even had some samples of the film. He gave them to me and I got down to work.

Soon, I learned that similar film was produced by the American company Pennwalt in the Pennsylvania town with the odd name King-of-Prussia. I called the company and ordered a good supply of the American-made film for my experiments. The film was very sensitive and easy to handle. I designed an electronic circuit to amplify and record its output. Ray and I built a crib mattress with the film placed inside of it. If one tapped lightly with a finger on the upper side of the mattress, the recorder produced a strong deflection. The mattress became a very sensitive transducer of variable pressure. When we tested it with a real baby, the oscilloscope screen clearly showed us the heartbeat, but very little breathing. The mattress was primarily sensitive to the vertical force, which was why we could see the heartbeat. But we needed to see the breathing, from which we got little response. I was discouraged and confused. The mattress was very sensitive, but not to what we wanted.

I turned to books on physiology and anatomy in the hope of finding some clue for the design of my noncontact respiration sensor. Thinking about it day and night, I came to focus my attention on the diaphragm, which in the living body moves back and forth, from the lungs to the abdomen. The diaphragm functions much like a piston in an air pump. During inhale, it pushes out, thus creating lower pressure inside the body, and the outside air rushes in. During exhale, it moves up, pushing the used air out. It was this that gave me the idea that when the diaphragm moves, the center of gravity of the lying infant's body should deflect as well. My mattress was sensitive to a vertical force, that is, to the baby's weight. Now, I realized that this weight was shifting slightly from head to toe with every breath. What I had to do, then, was to modify the mattress' transducer in such a way that it became sensitive to a horizontal displacement of the body weight.

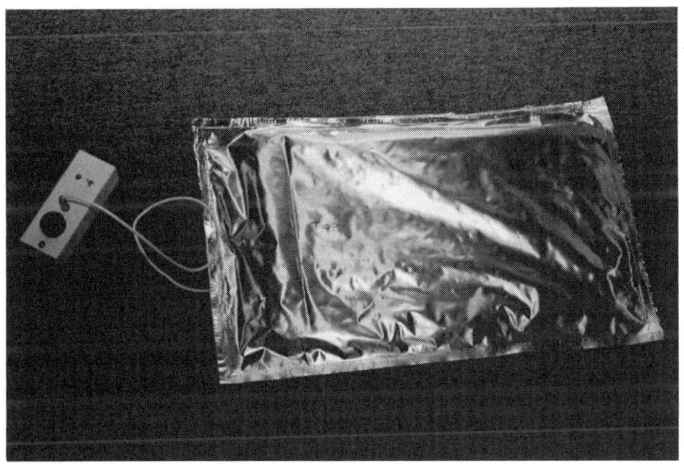

Sensor and monitor for detecting apnea

The solution I came up with was quite simple and at the same time original. I modified the sensor in such a ways that it became sensitive to force moving in a horizontal plane. With this, the mattress became sensitive to respiration and less sensitive to the beating heart. The solution proved to be both effective and novel. Timex applied for a patent which was granted without too much argument with the Patent Office examiner. It became my second U.S. patent.

Meanwhile, Timex was searching intensively for other ideas which could be developed into high-tech health-care products. One day Al Willis called me into his office for a meeting. In the office I found Al with our two marketing managers and one of our engineers, as well as two other people I had never met before: a short gray-haired gentleman (his name was Marthy K.) and a tall brunette woman in a garish green dress with a deep décolletage which concealed little but suggested a lot. She was Marthy's wife.

At first, I thought that Marthy was just nervous, but later I realized that he was a fairly progressed neurotic. A psychiatrist by education (what was it—an occupational disease?), he had decided to become a businessman. He had set up a small company in order to develop a machine for measuring blood pressure and hoped to sell the design to Timex. Such was the message he tried to convey to us, though not entirely successfully, as his wife had a mouth even larger than her décolletage. She kept interrupting everyone, talking fast and laughing loudly, and pretty soon had everyone feeling that she was the one running the show.

Al Willis asked me to evaluate the technical merits of Marthy's proposal. I was afraid that might be no easy task, since our visitors had no working prototype, just a few hand-drawn sketches and a verbal description. When I began asking the entrepreneur-psychiatrist a few questions about the operation of his device, it became apparent that either Marthy had not studied hard enough at medical school, or had effectively forgotten everything. His knowledge of the basics of hemodynamics was so primitive, his understanding of the methods of blood pressure measurement so naïve and his background in engineering so utterly nil, that before long it was clear to everyone in the room that his proposal was a hoax.

My questioning was apparently very upsetting to our visitors, for the brunette was soon shooting me furious looks (in spite of my appreciative glances at her endowments), while Marthy's hands began shaking noticeably. Suddenly, he jumped to his feet and started shouting in my face, using such filthy language that I realized that my English vocabulary was still quite undeveloped. After this verbal outburst from our guest, I had no choice but to stand up and declare that I felt that my presence in the room was no longer required, and took my leave. The meeting was wrapped up very quickly, and our emotionally disturbed visitors left.

Marthy and his wife visited Timex on a couple of other occasions after that, hovering around me with apologies and obsequious smiles. If he had been an inventor, I might have understood and forgotten his vulgar attack, for I know that great minds sometimes live in their own worlds and cannot be judged by ordinary standards. But he was neither an inventor nor a businessman. He was a gold-digger who saw an opportunity to sell some snake-oil. In effect, he had nothing to sell: there was no original design, no new methods of measuring— nothing. He was hoping to get from Timex a development and royalty contract. Then he planned to hire some engineers who could make a design for him which would produce a steady royalty flow. His perception of business was so primitive that he never got the idea that with no patent in his hands and no working prototype, he had nothing to sell.

However, one good thing came out of his visits. It made our people realize that we needed a good design for a blood-pressure machine for home use. To investigate the state of the art, we bought every available hospital blood-pressure monitor we could find and carefully analyzed their construction. The machines were all good, but adapting any of them for the consumer market would have been quite expensive. After giving it some thought, I said to my boss, "Al, give me one month, and I'll design for Timex a consumer blood-pressure machine."

He did not believe that I could do it that quickly, but agreed to give it a try. We got together a small team of engineers and technicians and set to work. Timex, with its enormous resources, was a great help. Not only did we have in the same large room with us a number of talented engineers whose advice was available to me at any time, the company was also equipped with modern machinery which we could use to construct the prototype. Timex's people, who traveled around the world, helped us to source many components which we needed for a cost-effective design. These included an arm cuff with a Velcro lock, a piezo-electric microphone, a display, a microprocessor, and so on.

On the other hand, being a large bureaucracy, Timex tended to turn its wheels too slowly at first, then too fast. It had a mentality typical of many large companies, according to which business should be marketing driven. I stress—not *market* driven, but *marketing* driven. In effect, the entire enterprise was controlled by the marketing department, who sincerely believed that they had the ultimate knowledge of what the consumers needed and how projects should be managed.

Many marketing people, even the best in the business, see themselves as masters, and the research, development, and other technical people—as slaves. The reason for this is obvious—marketing is the interface between the outside customers and the internal business structure. A great majority of marketing people, however, lack a good understanding of technology and fail to see the broader picture, unless they work in very close contact with the inventors and scientists. But that happens only very rarely. In most cases marketing managers perceive a fairly skewed picture. Usually, there is perpetual conflict between these two groups. Running an entrepreneurial business is not unlike the situation on a battlefield: the general is never positioned on the front line nor does he lead the attack. This would invite defeat. Instead, he is posted behind the troops with a clear view of the entire field. The Japanese understand this very well. Most of their more innovative companies are run by technology knowledgeable people who also have managerial skills. The consideration is simple. It is not a big deal for an engineer or scientist to acquire additional knowledge in marketing, sales, finance, and business administration. Professional marketers, salesmen, or accountants, on the other hand, cannot become technically literate to a sufficiently high degree. For them, this is just immeasurably more difficult.

No doubt, in some cases, it makes sense to allow marketing to run a small business, especially if it is not of innovative type. Yet I feel that allowing marketing or financial people to control a technology-oriented business is a gross oversight which hurts many companies in America.

To aid in developing the blood pressure monitor, Timex's marketing group hired an industrial designer, a packaging designer, ergonomic consultants, and other such specialists it deemed necessary to ensure the appeal of the blood-pressure machine to consumers. Overwhelmed with a sense of their own "importance," the marketing group never bothered to discuss their work either with me (I was, after all, the only biomedical engineer in the entire company) or even with our medical consultants. Allowing the Timex marketing people to run a medical enterprise was especially unwise in that the only background they had was in selling watches. They had no prior experience in dealing with the health care market for consumers.

After about four weeks, I had a bench prototype of the new blood-pressure machine ready. My left arm had taken on a bluish hue as I used it as a "guinea pig," taking my own blood pressure at least a hundred times a day. No decent simulator existed to replace a real human arm, so I kept the inflating cuff on my arm from morning till night. I soon discovered that my own blood pressure was very high and went to see a doctor. This probably saved my life, because it was then that I started taking medication to keep my hypertension under control.

It took another month to test the prototype on several hundred Timex employees, and by the spring of 1982 we were ready to build a pre-production prototype. I went to Al Willis and said that we knew exactly how the machine should be built and suggested that it was time to talk to an industrial designer and give him directions for the construction of the housing. Al replied that the marketing department was way ahead of me, and that the industrial design had already been done. The entire plastic box, including all controls, was complete, and we were anticipating the arrival of the parts from a molder any day now.

This news disturbed me greatly. How could anyone possibly design a shell without knowing what was supposed to be inside and how the thing was to be used? Al cracked an apologetic smile and said that he did not run the business, marketing did, and my job was to make sure that the guts fit into the ready-made belly.

A few days later, someone from marketing brought me the plastic parts for the prototype. They were not bad at all. The housing had been made by Laszlo Tapolcai, a Hungarian industrial designer (we later became good buddies). But of course, he had been working under the strict guidance of the marketing staff. Our technician Ray put the prototype together and calibrated its pressure sensor. Apparently, Laszlo knew what he was doing because everything fit inside nicely and looked neat.

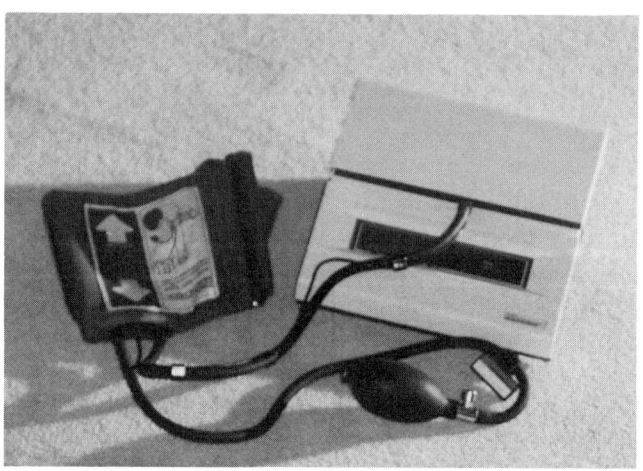

Timex blood pressure monitor

I took my blood-pressure. It was at this point that I noticed that the rubber tube connecting the cuff to the pressure sensor was plugged into the right-hand corner of the desktop-sized device, right above the display panel. As a rule, blood-pressure should be consistently taken from a patient's left arm. For someone to take his own blood pressure, he has to use his right hand to squeeze the bulb of the air pump. This design, with the pressure sensor at the right side of the machine, was very wrong. Not only did the rubber tube obscure the display, it was also too short to run from the right-hand corner to someone's left arm, inadvertently forcing one to put the cuff on the right arm and squeeze the bulb with the hand of that same arm. I called the vice president of marketing and told him that the design had be changed and why. The V.P. dispatched one of his men to investigate whether this was indeed important. I was then officially notified that switching the pressure sensor connector to the left side of the machine would require a total redesign of the product housing, and that it was both too late and too expensive to undertake this as the entire set of the hard tools had been already cut.

Designing that blood-pressure machine had been a tremendous challenge for me. When you develop a precision instrument or any new product, you try to select the best components to do the job, calculate all possible sources of error, run exhaustive testing under worst-case scenarios, and pay attention to myriad other equally important engineering details. This is the typical engineering routine. However, when you work on a consumer device, where cost is of the paramount importance, you must strike a balance between performance and the price for that performance. Often, you cannot afford to use what you would like and must resort to less reliable or less than perfectly accurate parts. My experience in designing medical electronic instruments in the Soviet Union was of great help. I knew how to get the best results using poor components. In short, I had more than a few engineering tricks up my sleeve. This was to help me a great deal, but nothing should be pushed beyond a certain limit. An engineer must never try to fool Mother Nature. If you are smart enough, you may play games with her or show her a magic trick or two, but never try to cheat her. She always strikes back, and she can hurt you badly, because Mother Nature never breaks her own laws.

Bruce Dodson, the marketing manager for medical projects, demanded that I do everything possible to save every single penny.

"Can you imagine," he used to say, "if we sell a million of these machines (and surely we are going to sell much more), every single penny you save will result in ten thousand dollars profit."

I reviewed my circuit again and again, and was able to find a number of interesting solutions to reduce the components count. One such solution was even filed with the Patent Office and became my third American patent. The marketing department employed an engineer named Jerry, whose job it was to serve as technical liaison between the R&D group and Bruce. He would form his own judgments, not always correct, and reported to his boss on what he thought I was doing either right or wrong.

One day, I got a call from Dodson, who said, "Jerry told me that you have an extra transistor in the power supply circuit. His opinion was that we don't need it. It would save us a nickel—a lot of money."

I explained to him that the power supply fed the entire circuit, and that the transistor, even if not specifically required for normal operation, reduced power consumption and tripled battery life.

"We don't care," was his response. "A consumer pays for the battery, but we pay for that extra transistor. Take it out." So I took it out.

Evidently, the marketing group was under internal political pressure to get the blood-pressure machine on the market as soon as possible. Besides the machine and my crib alarm, the company had two other consumer medical products under development. One was an electronic fever thermometer and the other was a bathroom scale. All these medical products were to be sold under the brand name "Healthcheck." None of the projects was ready for shipping, and to justify the very existence of the company and to protect their jobs, the people who ran the business wanted to put the product on the market which, they believed, was closest to a completion, and that was the blood pressure machine.

Production was subcontracted out in Singapore, and within a couple of months we received the first shipment of the devices. The Timex production group deserved a great deal of credit for implementing such quality production in so short a time. I was very excited, for it was my first invention to be produced on such a large scale.

The company pushed the blood-pressure machines through the same distribution channels it had used for so many years to sell the popular Timex watches, that is, through drug stores. The sales were terrific. In first six months alone, we sold several hundred thousand units. It was the first consumer electronic blood-pressure machine in America, and many people really needed it. The retail price was $70, no small sum, but quite affordable for those who must monitor their own blood-pressure daily. As I was well aware from personal experience, knowing one's blood-pressure can be a lifesaver. Everybody involved in the project was pleased. I too felt good about it and was proud of my contribution.

Several months later, however, the general excitement was fading. We experienced a very high return rate and received many customer complaints which the service department was unable to handle. The marketing department began to realize that manufacturing and selling the high-tech health care products was a totally different business from selling watches. There were a number of reasons for the setback. I remember three of them, in particular, as they related directly to the product design.

Number one was the absence of that damn transistor. With the machine's power consumption relatively high, consumers had to change the battery every couple of months. In doing so, they often mistakenly reversed for a moment the polarity of the battery contacts. Since there was no protective transistor at the input, even a fraction of a second was sufficient to zap the sensitive microchip in the circuit. The result was a damaged blood-pressure machine that needed to be sent back to the company for servicing. This problem affected, if I remember rightly, about 30% of all the units produced. Now, thirteen years later, I still use this same blood-pressure machine at home every day. It is a very good and accurate instrument and I am proud of it, but during that time I myself have twice had to replace a damaged chip. Surely I know how to put the battery in. Yet mistakes happen.

The second problem was with the rubber hose connected to the right side of the instrument. Since most consumers seem to have an aversion to reading instruction manuals (the old cliché: "if all else fails, then read the instructions"), they often placed the cuff on the right arm, assuming that since the hose was positioned on the right side of the machine, the cuff must go on the right as well. This resulted in a spurious readings of blood pressure and lowered confidence in the perceived accuracy of the device. Consumers complained and marketing panicked. To rectify the problem, I suggested producing an audio cassette with detailed instructions on the use of the machine to be enclosed in each box. My thinking was purely psychological. I felt that even people who never read manuals would still be curious to hear what was on the cassette. Hence, it might be a sure way to deliver the right message. As far as I recall, it was the only suggestion I ever made which was immediately and favorably accepted by the marketing group. However, they managed to screw it up very quickly. About two months later Bruce told me that the cassette was a great idea and that the consumers liked it.

"Wait a minute," I said. "You mean that you've already produced the cassettes and didn't even ask me to review the text? What if the message wasn't right?"

"We didn't have to show it to you. We know our business," answered the high-flown Bruce.

I ran to the customer service department and asked for a copy of the cassette, which I then took home and played on my stereo. Sure enough, the message contained several gross errors certain to make the situation even worst. And they did, as the following few months showed.

The third problem stemmed from insufficient testing of the circuit before the product was released. We had tested it on hundreds of Timex employees, but this was a relatively healthy population of a working-age men and women. Many of our real customers, however, were elderly retired people. We had never tested our algorithm on that age group, and the sensitivity of our microphone was not sufficient for detecting the Korotkoff sounds from human arms with low muscle tone. As a result, with a relatively large number of elderly patients, the blood-pressure machine was not entirely accurate. Clearly, I myself had to take the blame for that. I had never analyzed the potential customer base and thus had never realized that we needed to broaden the testing.

The company took several emergency measures to rectify the problems. The circuit was modified to include a protective resistor (which was not as good as a transistor for saving energy, but at least it would prevent the circuit from being zapped). The customer service department was given better instructions on how to handle the complaints, though the damage had already been done. At about this time, competing blood-pressure machines developed and produced in Asia were beginning to flood the market.

Meanwhile, the decision was made to develop a second-generation blood-pressure machine with an automatic air pump and a more sophisticated algorithm. I started working in that direction together with another engineer, Bruce Kamens, and a wonderful mechanical designer named Paul Wuthrich, a Swiss engineer who held nearly two hundred patents for various watch mechanisms.

Some people never learn from their mistakes. Once again, the marketing group presented us with a pre-built housing with the hose attached on the right-hand side. "For consistency," I was told. This should not have been as serious a problem as with the first machine, since the air pump was now automatic, but still, there would be too many people who would not use their left arms. In any event, we successfully completed the design and even built quite a few pilot units. However, Timex never marketed the new blood-pressure machine and later licensed it to the marketing manager Bruce and his orderly Jerry, who later, upon being laid off from Timex, started their own business to sell the devices. Just recently, ten years after I myself left from Timex, I learned that I share a patent with B. Kamens and P. Wuthrich for that automatic blood-pressure machine. Timex filed it after my departure and no one told me about it.

One day, I was summoned to the office of Al Willis, who informed me that the company had decided to terminate my crib alarm project. The problem was a fear that if we made and sold the alarm, sooner or later some unfortunate baby would die in a crib anyway. There was no doubt in minds of the company managers that, in this attorney-infested country of ours, a huge law suit would inevitably be brought against Timex Medical Products which would very likely kill the entire division and might even severely damage the parent company. Who can blame Timex for such a decision? Still, I was terribly disappointed to see three years of my work go nowhere. Yes, I received a patent, published a few papers, even presented the design at a couple of scientific symposia, but my real goal had been to come up with a product to save lives, not a pile of papers. Nevertheless, I shared Timex's concern and felt that if the decision had been mine, I would probably have done the same thing. Certainly it was not the only potentially good product to be killed for fear of the legal system.

Upon termination of the crib alarm project, I was able to spend all my time developing a new automatic blood-pressure monitor.

The group working on a fever thermometer was closing in. The marketing people had as their goal the development the fastest thermometer in the world—not a bad idea, given that all oral thermometers required several minutes to measure temperature accurately. There were also, of course, so-called predictive thermometers which used a special algorithm to shorten that time, but these thermometers were designed mainly for hospitals and the algorithm was patented by IVAC, a San Diego company. Timex's engineers decided to make a fast thermometer which would not require any special signal processing. For that, they needed a very fast probe which, when placed under the tongue, would very quickly heat up to body temperature.

To make a contact temperature probe that works faster, it must be very light and very tiny. The designers placed a sensitive element (called a thermistor) at the end of a thin-walled plastic straw. As with the blood-pressure machine, the thermometer was tested on hundreds of Timex employees. There was no doubt that the thermometer was indeed the fastest ever made. In a water bath it measured temperature in just one second, and within about ten seconds in the mouth. Everyone was very proud of the achievement, and production in Singapore was initiated.

When the first shipment arrived at Timex, the principal designer of the thermometer made me a present of one of the pilot units for my personal use at home. Just at that moment the telephone rang. It was my wife Irena, who told me that our two-year old daughter Julia was sick. Irena said it looked like the girl had a high fever, but that there was no way to measure the irritable child's temperature. It was impossible to force her to hold a thermometer's probe under her tongue for two minutes, nor would she consent to sticking it under her armpit for five.

I said, "That's great. I'll be bringing home the world fastest thermometer and we'll test both at once—the thermometer and Julia. Don't give her any medication yet!"

When I got home, I spoke intelligently to little Julia and convinced her to put the probe into her mouth for just a few seconds. The little one had already grown several brand-new teeth. After I gave her a small bribe, she opened her mouth and I stuck the probe in. Clang! She immediately snapped her teeth shut, biting off the tip of the thin-walled probe. Fortunately, she did not swallow it, and the thermistor was still hanging by two tiny wires. I carefully took the remains of the probe from her mouth, Irena gave her some Tylenol, and the next morning I brought the damaged gift back to its designer.

The poor guy turned pale. He grabbed the crippled thermometer and ran to the office of the V.P. In effect, Julia's careless biting forced Timex to do some major reshuffling in the design. The probe straw was switched to a more durable plastic, along with other substitutions. Unfortunately, all these modifications caused the thermometer to work much more slowly, and the fast speed, its main competitive advantage, was lost.

Timex was going through a difficult period. The 3-D camera project had been terminated. The personal computer business was going haywire. The crib alarm had been canceled. The bathroom scale and the thermometer had serious design flaws and were not ready for launching. The blood-pressure machine was still selling well, but suffered from an unacceptably high return rate. The company was losing a lot of money and massive lay-offs began. The first to go were the camera-related people, then the computer hardware and software engineers and technicians, then the medical products personnel. Pretty soon, I was the only one left in the whole of Timex Medical Products, Inc. The company was generous to its former employees. Those who were forced to go, received severance pay in proportion to the number of years they had served. I saw that there was no future for me at Timex and, as before, seriously considered starting my own business. My one concern was that I might not be laid off and would have to resign on my own. That, obviously, would prevent me from receiving any severance pay. Nevertheless, one day, to my great joy, I was summoned to the boss's office and given my pink slip. It was May of 1983.

Thinking back on my years with Timex, I have mixed feelings. It was a company blessed with entrepreneurial spirit. It had a group of very talented specialists. It identified several great opportunities. It had sufficient financial and human resources to put a number of potentially highly successful products on the market. Yet, despite all these advantages, things did not work out. In the end, Timex itself survived. It still makes best what it always made best—watches and clocks. But why did its other enterprises fail? Perhaps, someone better informed and more knowledgeable than I can give a better answer to that question, but here are my personal thoughts.

Probably, the first reason was of a general nature. I do not believe that a large company can efficiently do that which is essentially entrepreneurial by nature—the development of an entirely new product or process. It is an old piece of wisdom, almost a cliché, that a large business can not innovate. History, of course, knows some great inventions made within large corporations. A good example is the invention of a transistor by Bell Laboratories. However, Bell Labs were a very unusual entity within AT&T. Hardly any other company in our history, and virtually none today (probably even including AT&T itself), has allowed its scientists to devote themselves to basic research, spending endless time and millions of dollars on something exotic or highly speculative, which might never produce any positive cash flow.

Nevertheless, it is almost a law of nature that the enterprise that does not innovate inevitably ages, declines and may even die. In order to survive, it has no choice but to innovate, yet it must do so strictly along the lines of its primary business. For example, the best innovations for Timex are new watches and clocks, not computers or medical instruments. Side trips into other markets are never any good for the big guys. History provides plenty of examples of this. Take, for instance, the development of a commercial computer. The global computer enterprise began in 1949 with the invention of the transistor, and every large electrical company leapt onto the bandwagon: G.E., Westinghouse, RCA and so on. But every single one of them failed. And who won? The companies which were at that time either still small, like IBM, or non-existent.

The reason why a large company cannot innovate and the cause for Timex's failure to gain a foothold in other markets was that a large company is a large bureaucracy whose primary goal is to serve its own needs. Often, that goal is not to maximize profit, as many might think. The overriding task is rather biological, so to speak, not unlike that of a single-cell organism: to survive in a hostile environment.

The Soviet Union, where I spent my first life, was, in effect, a huge corporation, and while working for Timex, I could never get over its drastic resemblance to the Soviet Union. No enterprise, even in such a democratic country as the United States, can be a democracy. At best, it is an authoritarian entity, and often a dictatorship. The company's behavior tends to influence the behavior of its employees, who mimic their master to serve him better. It is indeed sad, but under such circumstances people's behavior often becomes hostile. There is a lot of truth in the saying, "It's jungle out there." The result of that survival tendency is a search for safer ways to live. These ways, however, are not compatible with entrepreneurial behavior. Employees in a large company are mostly concerned with preserving their own positions, or perhaps, with moving up the bureaucratic ladder. They rarely want to risk their careers by doing anything unconventional or risky. The reward is simply not there, whereas danger is.

There are also political reasons preventing large enterprises from being innovative. A peculiarity of the Soviet dictatorship lay in its fundamental adherence to ideology. Any ideology is a supreme thought, a mystical idea which needs no proof, merely unquestioning faith. Any doubting of that ideology was considered a heresy—a capital crime and subject to severe punishment. Though all that is in the past on a statewide scale, here in America I know of some large companies that have developed their own, smaller ideologies. As with any political ideology, they may sound very good and even progressive, for instance: "Quality first!" or "Our customer is our master." But that illustrious nobility is just a shell, with very little goodness beneath it. The problem is that an ideology does not really care about actual quality or real customers. It cares only about adherence to the slogan. And there is no room for innovation when people are concerned with serving slogans. This is why I never liked working for a large corporation, be it on the vast scale of the Soviet Union, or merely any of the major American companies. I simply do not fit into their structures. I have been there and I feel much better elsewhere.

Entire books can be written about why a small company can do such a better job in developing new ideas and new products. Much of it boils down to the fact that a small company has less red tape, is more dynamic, more focused, and (usually) less constrained by political factors.

Even if financed by outsiders, be they individuals or corporations owning substantial stock in the business, a small company is generally run by entrepreneurial shareholders possessing complementary talents: inventors, product developers, manufacturing experts, marketing professionals. Such small groups working together in concert are much more efficient in the business of making innovations. My experience led me to the conclusion, which is hardly original, that small companies do much better in inventing and developing new ideas, while large companies are much better in manufacturing and selling the products.

In a small company, the key players, who usually are the shareholders, are strongly motivated by the possibility of making it big. They are driven by a carrot, rather than by a stick. In a successful enterprise, the key players can fulfill their own personal goals, such as achieving a high-level position, making a lot of money, or just having fun. If their enterprise is successful, they are all successful, and that alone is a fundamental condition and driving force for achieving the goal.

I believe Timex could do much better if it would simply serve in the more limited capacity of a venture financier for small independent enterprises. Instead of having its own employees developing new products, it should set up several product development companies, one for each line—photographic, computer, medical, etc.—and give these companies complete freedom. A small dynamic company ran by shareholders-entrepreneurs could develop products faster, more cheaply, and with less risk of mistakes. True, mistakes happen, but a small company can handle them quickly and decisively. If these enterprises are successful, Timex could license these products from them for manufacturing and marketing. In exchange for its involvement, it could have the right of first refusal for any new products, giving it the first opportunity at an invention before it could be licensed to someone else.

If Timex acted as a customer rather than a master of such small businesses, it would be in a much stronger position. Everybody would make money in such a situation, and even if one of these companies failed, spreading the risk would definitely minimize the exposure of the financing party.

12
Mikie

> You will do foolish things,
> But do them with enthusiasm.
> -*Colette,* French novelist, to her daughter

He looked familiar the first time I saw him. The facial features, posture, body language, and even the manner of talking were similar to those of the TV celebrity Jackie Gleason. The man was grossly overweight, but very lively and nimble, like mercury. Energy burst from him, and his stocky figure was always in perpetual motion. He usually carried two enormous briefcases. One briefcase contained business papers, technical literature, data sheets, and samples of various electronic components. The other was full of presents. He always had some sort of souvenir for you from the Orient: a folded Chinese paper lantern with red and gold ornaments, the latest Japanese electronic gadget, a fountain pen, or just a postcard with a breathtaking skyline of Hong Kong or Tokyo. He possessed a cheerful nature, was very sociable and well mannered. He loved Oriental women, a glass of dry sherry, and a large plateful of sashimi.

His name was Michael, but he preferred to be called Mikie. He spoke a beautiful King's English, betraying his British origin. He was a handy man: he could wield any tool with the skill of a professional carpenter, welder, mason, or painter. He was always ready to help whenever you needed him. He was someone you liked immediately and always wanted around.

We knew each other for about three years, and when he disappeared, I badly wanted him back. Today, I am not the only one still looking for Mikie. He is wanted by law enforcement agencies in the United States, Canada, the United Kingdom, Korea, Hong Kong, and a number of other countries. Interpol has chased him across the globe, and I hope that one day he will be put behind bars for a long time—because Mikie was a high-caliber crook and a shameless thief. But when we met, I knew nothing about his second nature and believed that I found a good and caring friend.

I was looking for financial support for several of my inventions when, among a few other international wheeler-dealers, Mikie walked into my office. We talked for bit, chatting about my various projects. I showed him some prototypes, including my infrared motion detector. I was particularly proud of that invention, because it was a significant improvement over my old light switch, for which I was just beginning to receive royalty payments. The new sensor could operate in total darkness and had a broad operating range, but its most attractive feature was its very small size, being no bigger than a ping-pong ball. At the time, the only light switch on the market that could turn off the lights in an unoccupied room was one based on my 1978 design, which was licensed to the Illinois company Intermatic. My newly invented infrared sensor was suitable for use inside a small wall switch, which made it an enticing option for automatic lighting fixtures.

When Mikie saw the prototype, he said that it made sense to show it to a few manufacturers in the Far East, where he was heading the very next day. He suggested that I give him the unit and a short description, and he would see what he could do. It seemed a practical business proposal and I saw no reason not to agree.

Mikie called me a week later from Hong Kong. He was practically shouting with excitement, "Listen, Jake, the sensor is a great success!!! I've demonstrated it to several Oriental manufacturers and there's been immediate interest. They all want it! But listen to this, I am *not* going to give it to anyone! Yes, you got it right—to no one! Don't interrupt me, I know what

I'm doing! I'll be back to Connecticut next Monday and I'll tell you about my idea. You just wait, my friend!!!"

The following Monday he appeared at my office as promised. He told me that seeing the enthusiastic response to the sensor among so many Asian manufacturers had made him think that, in fact, he should not arrange a licensing agreement for me with any of them. Instead, he asked me to license the invention to him, Michael Coveley. He explained that he had always wanted to work for himself and regarded my infrared sensor as the opportunity of a lifetime.

Since at that time I saw no way to do anything else with that patent, I asked him to describe what exactly he had in mind. It turned out that he had not slept at all during the fifteen-hour flight from Tokyo to New York. He had spent the time devising a detailed business plan for starting up a new company. Its purpose would be to develop, manufacture, and sell my new motion sensors for a variety of customers ranging from lighting manufacturers to security industries. He offered me a very generous royalty, but no down payment, saying he wanted to invest every penny he had into the new business. The only thing he wanted from me was to provide any technical help he might need.

Needless to say, I was delighted. I had spent so many years looking for a business partner, and here, out of the blue, someone was offering to acquire a license from me and even wanted to start a new business around one of my inventions. Moreover, I would not even have to participate personally and could continue working on my other projects without interference or interruption. All I would have to do would be to give him technical advice now and then. Of course, I agreed. We shook hands and went to see a lawyer who helped Mikie incorporate his new business and sealed a licensing agreement between me and the newborn company.

Mikie rolled up his sleeves and began working very energetically. First, he quit the job he had with a Midwestern company.[1] Then, he leased a space in a run-down building. With his own hands he remodeled it beautifully, buying furniture and office equipment and engaging several subcontractors to fashion the special production tools he needed to manufacture the parts for the motion detector.

This incredible haste bothered me a bit, because I knew very well that there was still a long way to go from the prototype I had made to fashioning production tools. I told him that before setting up the production facilities, he would do better to concentrate on the development process. He needed to engage development engineers and technicians to finalize the design and do thorough testing under various environmental conditions. That could easily take at least six months to a year, and only after all the designs and trials were successfully completed should he worry about fabricating the production tools. Apparently, he had different ideas.

"Jake, my friend, I don't tell you how to do your job, so let me run my business the way I want. Believe me, I'm good at that, and you'll be pleased with the results."

Even if I disagreed with him, he had a point. After all, it was his business, he had invested his own time and money in it, while I had merely licensed my patent to him. Fair was fair.

When most of the tools were ready, and it looked to me that I would soon see how all those small parts fit together, Mikie told me that he had changed his plans. He no longer wanted to run the business in America. He said that he had a much better idea. He took me to a Japanese restaurant for lunch, and once the raw fish was happily swimming about in my sake-filled stomach, he said, "It's very expensive, almost sui-

---

[1] Later on, that company hired a private detective to track down Mikie. Apparently, before leaving, he cheated them out of tens of thousands of dollars through credit card fraud.

cidal, to run a start-up business in America. The government doesn't care if you succeed or fail, they just want to tax you. Bloodsuckers... This country has lost its entrepreneurial spirit and courage. Nobody wants to produce anything. Everyone wants to be a middleman and make dough from transactions. Or invest in real estate. The business climate for entrepreneurs over here is very unhealthy. I can't afford to make all these sensors in America—too expensive, too much red tape, too much risk. The only way to go is to move all my operations abroad, to some place where the government helps small start-up businesses."

"What are you talking about?" I said, baffled. "You have a family here, a wife and four children. Besides, if you go to another country, how can I help you with all the endless technical issues you'll have?"

"Korea!" His eyes were beaming. "It's a land of golden opportunities. If you go to a rural area there, like Yosu, the government gives you all sorts of grants to cover your workers' wages, free rent for your facilities, and a 'tax holiday.' Besides, Koreans are bright and hard-working people. I like them a lot. And don't you worry about my wife, she's quite used to my long trips. I've lived like that for nearly ten years now, and she doesn't mind."

"But what about the technical side? You can't do it without my supervision. The sensor is too tricky and you won't find any local engineers who can help you. I can't be of any help to you if you're on the other side of the globe. I'm too busy with my other projects. I can't travel to Korea."

"The telephone, Jake—the telephone and the fax! We live in the twentieth century! Here in America, I'd be bankrupt in a couple of months, but in Korea I'm gonna win with flying colors! We'll both win, trust me!"

I shook my head. I was torn between two contradictory feelings. On the one hand, he still had not convinced me that moving to Korea was the only reasonable solution for his financial difficulties. On the other hand, I admired his self-sacrificial willingness to relocate to the Orient, leaving his wife, his son and his three daughters behind. All that—just to produce my sensor. There seemed almost something heroic about it. What would you have thought?

Mikie organized a farewell party at his home in South Windsor, Connecticut. Apparently, he did not have many friends or even acquaintances. Besides my wife and me, there was only one other couple present. We sat on his backyard deck sipping beer from huge mugs and puffing away on the pipes that Mikie gave us as presents. The party was a bit tense. His wife did not look happy at all. She hardly spoke a word to us, and we had the feeling that she would just as soon have seen us all go to hell along with her none-too-beloved husband.

Mikie's luggage was skillfully packed in nearly a dozen huge trunks. He was taking everything imaginable with him, from personal belongings to fax machines and a number of power tools. The following morning he flew to Seoul.

He called me from Yosu. The trip had been very difficult: a powerful typhoon delayed the flight, he twisted his ankle carrying the heavy luggage, and the roads in southern Korea were flooded after the torrential rains. Nevertheless, he already had an apartment and the local authorities had helped him find a building for his factory. A couple of weeks later, I received a package from him containing color photographs of the freshly remodeled but still empty rooms of the spacious facilities located in a three-story building.

He continued to send me reports on his progress every three days or so. Each new letter contained more pictures with the same facilities, now filled with benches, shelves, and various new and used electronic assembly equipment. The walls were decorated with traditional Korean ornaments. Mikie also sent me pictures of his two dozen new employees, including a personal translator and security guards. To buy the equipment and recruit the engineers, he had traveled all through Korea and as far as Hong Kong.

Things went on like that for about four months, when early one morning I received a telephone call from Hong Kong. A lady with a British accent asked me if knew how to find Michael Coveley. She said that he was an old acquaintance of hers, but that she had lost track of him some years ago. She explained that she had gotten my telephone number from his former employer in the Midwest, who, she said, knew that Michael had some kind of business relationship with me. Thinking that Mikie would be delighted to hear from his old acquaintance, I gave her his address and telephone number in Korea.

Two days later, Mikie called me in the middle of the night. He was furious and screamed so loudly that I was afraid he might wake up the whole neighborhood.

"You are an idiot!! Who allowed you to give my address and telephone to anyone? I am here working my ass off, but you don't bloody care!! Do you?! Instead of helping, you are sending very disturbing people to me! Very disturbing!!"

I was at a loss. I could not understand what the problem was or why he was talking to me like that. What was wrong with giving out his address? Why should his whereabouts be kept confidential? I shouted back at him, which seemed to calm him down a bit. He began asking me every minute detail of my conversation with that woman. He insisted that I recall the exact words she and I had used in our conversation. Then he said, "Please, for Christ's sake, never, ever give my address or telephone number to anyone. If someone comes looking for me, just say that you don't know me. Is that understood?"

I was very disturbed with his behavior, but promised not to talk about him with anybody. When I asked what all the commotion was about, he replied, "It's just an old story, you know what I mean? A woman, a bloody jealous husband. He's been chasing me everywhere. I'll tell you all about it one day, but for now keep your mouth shut, will you?"

I thought that the incident was resolved and everything settled, but in another week Mikie called me again—this time from Taiwan.

"Jake, I need your advice. The situation is very disturbing and even dangerous. Have you seen the television?"

I said yes, that I had seen the news reports about the clashes in Seoul between rioting students and police.

Mike said, "Seoul is nothing comparing with Yosu. It's bloody hell over there. They don't show it on the news because it's a rural area, but believe me, it's very, very bad. My life was in danger. Nobody's working, there's rioting in all the streets, they throw stones, break windows, burn tires. Much of my equipment has been vandalized, the building is damaged. I've even had to flee to Taiwan. What shall I do? What do you think?"

I told him he should wait for things to calm down, then return to Yosu and see if he could continue. I did not believe that would be possible for a long time. But apparently Mikie had a different plan.

"Just yesterday I met a man here from the British consulate. He told me about the wonderful opportunities in England. The British government gives very good incentives to entrepreneurs who open businesses in economically depressed areas. Why not move my business to England? It's very sentimental for me. I was born in London, you know. Besides, there is no language barrier. What do you think?"

I had no idea what to say. "Look, Mikie, it's your life, your money, and your business. I don't know the local situation, either in Korea or in England. But of course, I think it would be much easier for you in the U.K. At least there should be no cultural problems..."

The next call I received was from London.

"Jake, my friend, everything is great! Of course, I had to leave all my equipment and even personal possessions in Korea, but that's nothing. I'll get much better stuff here. Yes, I lost a lot of money, but it doesn't matter! I am going to open the business for making your sensors in the North of England. I looked as far as Edinburgh, but finally I found what I need. There is a nice town called Scanthorpe, a very depressed area. They used to have huge steel mills there, but now there's nothing. The British government gives great incentives: grants, interest free loans, a tax holiday—everything I could dream of. I am moving there tomorrow. Oh, yes, one more thing. I got married in Korea. My wife is young and beautiful. She is with me and her name is Sol-Byung. Honey, say hello to Uncle Jakie."

This was quite a piece of news. I tried asking him when he managed to get a divorce from his Connecticut wife, but he just brushed my question off. Now I suspect that he never even bothered divorcing his American wife of twenty years. No doubt, he thought that it was perfectly fine to have separate spouses in different countries, as long as they never met.

He settled down in a town near Scanthorpe and, as he had done earlier in the United States and Korea, rolled up his sleeves and started building a company from the ground up. Before long, he asked me to come over and give him some technical guidance. I flew to London, took a train from Kings Cross to Doncaster, and from there another train to Scanthorpe. Mikie met me at the station and we drove straight to the factory.

It was very impressive. The building was brand new, with at least twenty thousand square feet. Mikie had already hired about a dozen employees, a very disciplined and hard working group. He himself worked at least twelve hours a day and never shied away from any job, either in the office or on the production floor. Everything was coming together very nicely, both the factory and the product.

As before, I brought his attention to the fact that the sensor was still quite immature. Some serious engineering work was needed before production could start. It was too risky, I said, to put the carriage before the horse. When I told him this, Mikie introduced to me two young fellows. One was an electrical engineer who had just recently relocated from London, and the other was a technician. Their job was to finalize the design of the sensor. I spent a day priming them on the design, though with little success. I told them what needed to be done, how to run the tests and what kind of components they should use. One of the difficulties they faced was finding a local supplier for the so-called plastic Fresnel lenses. I promised to look about in the United States for an appropriate vendor.

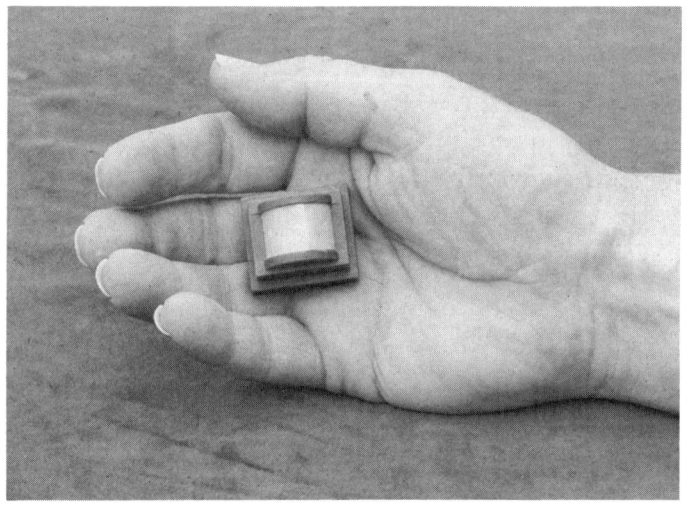

Infrared Motion Detector produced by Mikie's company

To celebrate my visit to the plant, Mikie organized a grandiose banquet, to which he invited some local authorities, his solicitor (a British lawyer) and accountant and their wives, and a couple of key employees. It was at the banquet that I met his young Korean wife. She was indeed very young (no more than twenty-one or twenty-two) and very beautiful, just like something from an Oriental calendar. The dinner was excellent, with plenty of British ale and French wine. The food, to my amazement, was also delicious, which was a pleasant surprise, because I heard many times before that "the English like tasteless food." Mikie, a skilled party man, made friends with the entire kitchen staff. Everyone, from the chef to the janitor, received presents from him, and we guests were rewarded with the tremendous variety and quality of the food and the royal service.

Returning to the United States, I looked for a supplier for the plastic lenses Mikie needed to make the sensors. The vendor was a small company in Fort Worth, Texas, called Fresnel Devices. I called Mikie to let him know that its president, Dr. Dick Klaytor, planned to be in Connecticut in the near future. Mikie said that it was very important for him to meet the vendor and decided to come over as well. He flew to New York and spent only one week in America, but in that time accomplished a great deal: he met with Dr. Klaytor, visited another vendor, the Pennwalt Corporation near Philadelphia, sold his house in Connecticut, opened a U.S. bank account for his business to pay the bills to the American vendors, visited his children and former wife in Florida (did she know that she was "former"?), and before he left again for England, we had a terrific sushi lunch in Manhattan.

Mikie told me that he had applied for more grants and loans in England, and that he also wanted to sell stock in his business. He proposed that I buy some. Just a week before this, I had received $20,000 in royalties for my wall switch. I thought that it seemed like a good idea for one invention to feed another. So I was foolish enough to invest all that money in Mikie's business. I believed that he was building a company with a great future.

After my first visit, I traveled to England a few more times. Sol-Byung gave birth to a charming boy whom they named Luke. Mikie was a loving father. He bought a house, and from that time on I stayed with them rather than at a local hotel. The new Coveley family radiated love and peace, and I thought that he had decided to settle in England for good. He had told me that he had been born in London in 1939. His mother was killed by a German bomb when he was only a year old, and he was brought up by his older sister. As a young man, he moved to Canada, and then married in America, where he had lived for the last twenty-odd years, counting, of course, his long trips to the Orient. And now, he was back in England.

Mikie was a grand master at building good-looking businesses. The company was sparkling clean, all the workers wore white coats, the walls were covered with beautifully drawn diagrams reflecting the engineering process and quality control steps. And I even saw some real progress in the improvement of the product. Production was escalating, Mikie bought some new sophisticated pick-and-place machines for the electronic assembly, and his warehouse was filled with thousands of the newly produced sensors. He even began producing various consumer products along with them, like personal security alarms called *The Cyclops* and outdoor lighting fixtures. Everything was great except one thing—the company was making very little sales.

"Mikie," I said to him, very discouraged when he refused to show me his sales reports, "why all these excessive measures: the super-clean facilities, expensive equipment, white coats, the warehouse with boxes up to the ceiling? You really don't need all this to make the sensors. Why are you focusing on all these bells and whistles? And where are your sales? I hope to start receiving my royalties some day, but so far I don't see any products moving. It doesn't look right to me. Why?"

"Hollywood, my friend, Hollywood! Nearly every day I have visitors from the banks and government. They want to see where their money goes, so I have to put on that show. I need your help with that, otherwise—no money, no business. To get all these loans and grants, I have to make it look expensive and efficient, very high-tech! You see, it's like a chain—I make a little Hollywood, that brings me more loans and grants, after that I expand the business, and then—your dream will come true. And you are wrong that I have no sales—I sell a lot, but so far mostly to Europe. That's why all the books are there, in my other office in Frankfurt. One day we'll go to Germany together and you'll see for yourself. Trust me."

He took me out to lunch at a country-style restaurant, and once there, continued, "But there's more to it. I'll tell you a secret. Soon, I am going to meet with a man who wants to... buy this company."

"What do you mean? Who would want a company in such an immature state, and why? And why do you want to sell it?"

"Because I want to recover my investments, that's why! The business is for sale. This man I'm talking about owns several businesses and is looking to buy more. You see, that's his game. He can buy my business relatively cheap, then he will invest heavily in it, beef up production, bring the value up, and resell it to another buyer or put public. It's called a leveraged buy-out. You know, it's a standard thing."

"What are you going to get from it?"

"I'll get a fat lump payment and a contract for a few years for me, because I'll still be running the business. And for you, I will arrange a lavish consulting contract, so you'll start making some money, too. And then, you'll begin getting your royalties. It's in your own interest, so you must help me. Without you, I can't make him a good technical presentation. And that's vital. Please, be here next month. I need you."

I had no idea what to say. Of course, I had heard many stories about acquisitions and bringing stock values up. I was not an expert in such financial games. I said, "Okay. As long as it won't take more than two days of my time and you pay for the round-trip ticket, I will be here." Mikie was very pleased.

It was time for me to return to London. Mikie said he would take me to the Scanthorpe station to catch the eight o'clock train. I had to spend the night in London because my flight back home was early the next morning from Heathrow. I said no, the eight o'clock train would get me to London too late at night, and I wanted to take a walk in the City. London is one of my favorite places to visit. I must confess that one of the enticements in visiting Mikie was the opportunity to see London again and take in a play at one of its theaters, or just walk around the City.

Mikie's secretary, however, warned me that it might not be wise to take a six o'clock train. "You see, on the Doncaster train there will be a lot of football fans coming from the match. Those guys are nuts. You really don't want to be in the same train with them."

Quite the opposite! I had heard a lot about the crazy British soccer fans and did not want to miss the opportunity of seeing them in person. I said to her, "In that case, it's exactly what I want. Get me a ticket and don't worry. I'll manage somehow."

She was right—the train was a real zoo when I boarded at Doncaster. The car was packed with thoroughly intoxicated fans all loudly shouting and running up and down the aisle, swinging their fists and whacking each other with their feet. It looked as though they were trying to reenact the game they had just seen and resolve some disputed issues. That was my best guess, for I could not understand a word they were screaming. Professor Higgins in the George Bernard Shaw play *Pygmalion*, could, depending on the accent, locate a speaker's place residence within a two-mile radius from central London. Probably, I was still too far from London, because I was not even sure it was English they were speaking, or rather shouting. I have always felt at home in England, except for some difficulties in understanding what people are actually saying. Indeed, I very much agree with the old saying that the English and the Americans are a single nation divided by a common language.

Two particularly excited soccer fans, unable to resolve their dispute, began demanding arbitration from the other passengers. The atmosphere became very thick and explosive. I was the only person in the entire car who did not participate in the quarrel. Eventually, this fact caught their attention and they decided to ask my opinion, or at least such was my guess. Since it seemed unwise for me to take anyone's side, I replied, "I don't speak English."

The fans were stunned—how on earth could somebody not speak their tongue? (It was English, after all!) This puzzled them so much that they cooled down for a while and left me alone, and I reached London without any problem.

Just ten days later, I flew to England again. This time, I was not going to Scanthorpe. Instead, I was to be met at Heathrow airport by Lui Rodriguez, a young investor allegedly interested in buying Mikie's business. We were supposed to meet at the airport cafeteria, and after an hour or so, take a small plane to

Jersey, one of the British Channel Islands located near the French coast.

Mikie was to fly directly to Jersey from Birmingham and meet us at the local airport. He told me on the telephone that the young investor was an international financial dealer whose family had acquired a substantial fortune in the South African diamond mines. He said that Mr. Rodriguez had been introduced to him by a local banker, that they had already had a preliminary discussion and were planning to meet in Jersey to negotiate the final deal. Jersey had been selected because it was a tax-free zone and Mikie wanted to open a bank account there to deposit the money he received after the supposedly hefty transaction. Lui was apparently a great expert at shifting money among international accounts, and had promised to help Mikie with those arrangements.

When I arrived at Heathrow around ten o'clock that morning, I checked my luggage for the flight to Jersey and went straight to the cafeteria. Our departure to Jersey was scheduled for one in the afternoon, so I still had time to have lunch and meet the young investor. I found a table on the second floor with a good view of the entrance and most of the departure hall. I ordered tea and began to wait.

About fifteen minutes later I spotted a young man with a dark mustache carrying a huge briefcase up the stairs toward the cafeteria. Apparently, he recognized me, and after a brief introduction he said in a slight accent, which I recognized as South African, "Look, in five minutes I should meet someone here at the airport. I'll leave my bag with you. The man I am meeting will bring me a suitcase with cash. A big one: exactly fifty thousand pounds, which I'll take with me to Jersey. It'll be just a few minutes. Please order for me... well..."—he looked at the menu—"this salad. I'll be back very shortly. We have plenty of time before our departure." With that, he ran downstairs and I looked at my watch. Almost noon. An hour until our flight, and the gate was only five minutes away.

"But what a strange man." I thought. "He's just seen me for the first time and is already talking openly about fifty thousand British pounds he wants to carry on the plane in a suitcase. What is this? Who in this day and age performs large transactions in cash, unless it's some kind of drug deal or money laundering? And is it legal in Britain to take that amount of cash out of the country? I should be careful with this man."

I looked at his briefcase. It was quite shabby and tightly packed. I pushed it under the table and sipped my tea, British style, for another fifteen minutes. Lui still did not show up.

When there were only thirty minutes left until our departure and Lui's salad had withered, I became a bit nervous. Where he was? How long could it take to meet someone? If his rendezvous had not shown up, at least he should have come back to tell me. What if this was some kind of criminal activity and Mr. Rodriguez had been kidnapped? These and other ideas ran through my head as I sat there waiting, fidgeting in my seat. With only ten minutes left before our departure, I heard my name being paged over the loudspeaker. I had checked my luggage and the plane to Jersey could not leave without me or until my garment bag had been taken off. At that moment I was positive that Lui had disappeared and I decided to run and board the plane alone. There was no point in my staying at Heathrow when Mikie was waiting for us in Jersey.

But as I stood up and started toward the cafeteria exit, I was stopped by the waitress, "Mister, you forgot your briefcase!"

I explained to her that it was not mine and that the man who owned it would probably show up soon to pick it up. But she said it was against regulations to leave any luggage unattended and she would call security if I left it there. I had only two minutes before my departure, so I grabbed Lui's briefcase and ran like hell to the gate.

They were holding the plane for me. To make sure that I was not taking anything dangerous on the plane with me, I opened the briefcase and quickly checked its contents. Inside there were several notebooks, a great many manila folders, a calculator, a small cellular phone, and a few miniature cardboard boxes with what seemed to be fairly inexpensive jewelry. Just as I was entering the plane, I saw Lui running toward the gate, shouting loudly, "Wait! Wait for me!!!"

As soon we were on, the plane took off. A few minutes later, after catching his breath, Lui began swearing and cursing in several languages, which besides English were (as I found out later) Afrikaans, Spanish, and Portuguese. And perhaps Zulu as well, though I am not sure. It turned out that the man he was supposed to meet had in fact shown up, but instead of fifty thousand pounds he had brought Lui only five. Lui was furious. He took a thick pack of bank notes from his pocket, apparently those five thousand pounds, and thrust it in front of my face with indignation. He ordered a glass of cognac from a stewardess, downed it in a single gulp, the way Russians drink vodka, and started cursing again.

"I'll kill that S.O.B.! I lent him money for one month to save his fucking business, and he was supposed to pay me back two weeks ago. He didn't then and promised today, but again nothing! Only these bloody five grand!" He again took the wad of cash from his pocket and waved it over his head.

I had no idea what to think of all this. Who was this Lui? Who was the man he was meeting at the airport? What was all this commotion with the money? Loan sharking? And what I was doing here in the middle of this bizarre mess? But whatever it might be, I saw it as just another adventure in which I had the fun of being, if not a participant (luckily), at least an observer.

Once the cognac began taking its effect, Lui relaxed and we talked. He told me that he came from a rich South African family of Portuguese origin. He showed me photographs of his beautiful and still young mother and even more stunning sister. His trade was international finance, and he specialized in arranging loans to the governments of small and medium-sized countries. He said that he was always looking for opportunities to park his huge commissions somewhere, and that Mikie's business seemed to him to be a sound idea.

Mikie and his solicitor[1] met us at the St. Hélier airport. They had a rental car and took us first to the best hotel in town, and then to a small cozy restaurant. Mikie paid for our seafood dinner, which was quite exotic. Lui lectured us on the art of financial deals. Most of what he said was new to me, but one thing was obvious: Lui was a grand master of financial manipulations on a global scale. He told us how he could predict fluctuations in the stock market and shifts in international exchange rates. He even claimed to be able predict the change in value of any currency immediately following any meaningful international event.

Mikie (right), Lui (center), and author in Jersey

After dinner, the four of us went to Mikie's room at the hotel, and the negotiations got under way. Actually, Mikie and Lui did the negotiating, while the solicitor and I were essentially passive observers, sitting on the couch sipping our thick ale and listening. I spoke only when Lui asked me questions about the technology, its trends, the market potential, possible future products, and the like. Lui said that he wanted to buy

---

[1] In England, there are two levels of lawyers, the higher being a barrister and the lower a solicitor. These fellows probably feel rather resentful when they visit America and see signs on people's doors reading: "No solicitors!" Wouldn't it be great if in this county we replaced this with "No lawyers!"

80% of the company and insisted that Mikie retain the rest to assure that, as a shareholder, he kept working there as hard as before. As far as I remember, the price was somewhere around two million pounds. Lui said that he would send his accountants to the company within the next few weeks to conduct an audit, and that everything could be wrapped up soon after that. Mikie was delighted, Lui was very business-like and the solicitor was very professional. And as for me? I just had fun.

It was almost midnight when Mikie, the solicitor, and I went down to the beach. Lui said that he needed to call someone in Miami and stayed in his room. A cold November breeze was blowing in from the Channel and the full moon had brought in the high tide. In Jersey the tides are of enormous magnitude, and we were intrigued to see how the level of the sea had risen so high that it completely changed the landscape. After strolling along the beach for a while, we returned to the hotel lobby. There we heard on the radio that Margaret Thatcher had resigned and John Major was the new Prime Minister of Great Britain. For Mikie and his solicitor, as British nationals, this was important news. I suggested that we see how well Lui did predicting the reaction of the stock markets following this news. Everybody agreed, and when Lui came downstairs a half hour later, we asked him, "Lui, we want to ask you a purely hypothetical question. What do you think: how would the stock market and international currency exchange react it Margaret Thatcher has resigned?"

Lui scratched his head and, as if reading from a computer screen, gave us the figures of his prediction. We wrote them down on a piece of paper and then told him why we had asked. The next morning we stopped at the newsstand to buy the issue of *The Financial Times*. We were quite amazed to discover that Lui's predictions were nearly a hundred percent accurate. The fellow surely knew his business. But just what was his business?

Once the talks in Jersey were completed, we flew back to England and I returned home to America. One interesting detail: as we were about to board the plane in Jersey, Mikie opened his briefcase and I noticed that he was carrying three passports. One was American, one British and the third Canadian. When I asked him why he had three passports, Mikie said, "I am British by birth, I lived in Canada for many years and became a Canadian citizen, and then I became an American citizen through my first marriage. I use the one depending on circumstances." I expressed my surprise that it could be legal to have passports from different countries, but he told me that such was the case (later, I learned that it was, in fact, illegal).

A couple of weeks or so later, Mikie called to tell me that Lui's auditor had come to the plant and audited all the books and records. I had no idea what he found out or what happened at the company after his visit, but when I tried calling Mikie's home a week later, the British operator told me that the telephone had been disconnected. I then called the company, where a strange voice answered the telephone. The man who picked up the phone was very inquisitive. Instead of answering my questions, he asked me who I was and what I wanted. He said that Mikie was not available to talk to me. When I called again in a few days, the operator told me that the company's telephone had been disconnected as well. I sent several letters to Mikie and the company. None of them were returned to me and I received any response from anyone. Mikie and his business had vanished.

A few months later, early in the morning, my telephone rang. It was Mikie.

"Jake, my dear friend," he said with a trembling voice, "everything is gone. Gone... They took my business, my home. I had to flee..."

"What are you talking about?" I was totally taken aback. "Who took your business and why?"

"The bloody government, that's who. I was late with the loan payments and they put the company in receivership."

"What about Lui? What about your agreement with him? Could you use his money to make the payment? Where is your wife and the baby, and where are you now? I have a lot of questions for you..."

Mikie told me that it all had to do with something mysterious about Lui. After his auditor and solicitor visited the company, Lui himself disappeared and Mikie could not get in touch with him, but one day the government people came and took over the business. He said that Sol-Byung and the baby had gone back to her parents in Korea and he himself had to flee.

"But where are you now? Where are you calling from?"

"Well, I'm in... a... in Barcelona."

"Where are you staying? Give me your number."

"You see, right now I am moving from this hotel to another one. So I don't have a number yet. You know, it's better if I call you."

Apparently, he did not wish to tell me where he was. I would not be surprised if he was calling me not from Spain, but rather from hell itself. I realized that a great deal of what he was telling me was a lie.

Then, suddenly, he said, "Jake, I am sorry that it didn't work out, either for me or for you. But look, I know you have a few more patents which you haven't put to any use yet. I want to start all over again, here in Spain. Or perhaps in Taiwan. Yeah... It's better in Taiwan. Give me another chance. And another patent. This time we are going to make it for sure..."

"Mikie," I said, "do you know the old saying 'fool me once—shame on you, fool me twice—shame on me'? I trusted you, but as far as I'm concerned, you did everything wrong and, I dare say, in a less than honest way. I was a fool to believe everything you told me. Where is the money I invested with you? Do you realize how much time I spent helping you? You want another invention? Okay, I'll give it to you. But not until you make me some money with the first one. Fix it, then we'll talk about the second one."

"Jake, how you can say that? I thought we were friends!"

We went back and forth like that for a couple of minutes, then he said that he had to go and would be calling me back soon. He never did.

Some years have passed since our last conversation. I have no idea where he is now and what happened to him, or to his beautiful Korean wife and the baby Luke. Mikie managed to steal a great sum of money, much of which his company had received as grants and loans from the British government and various banks. The money that I and some other small investors paid for his stock ended up in his pocket as well. He took the money and ran.

During these years I have received several calls from British and American private investigators looking for him. One time, Interpol even called me about Mikie. I have learned that his schemes and credit card frauds have cost several American and Oriental companies hundreds of thousands of dollars. A great many people have looked for him, all, apparently, with little success. That nimble fat man knew how to cover his tracks. I cannot forgive myself for being so naïve as to trust that lovable man and consider him a friend...

Thinking about him now, I believe that Mikie was a skilled con-artist of international stature. No doubt his scheme from the very beginning, was to extract money from governments and banks by posing as a successful entrepreneur. He used me as a front to gain credibility for his projects. His technical expertise was very rudimentary in spite of his sharp and lively mind. He needed me, my advice, and my patent to present himself in a better light and look like a professional high-tech start-up business. He masterfully built a company which produced a lot but sold nothing. To all appearance, that company was just a front for his illegal operations.

He had not wanted to remain in America and moved to Korea, because, I believe, he was on the run in this country. Likewise he had then to flee Korea not because of the student riots, but because there too he had smelled a rat. The earth was burning under his feet. He chose a small remote town in England, hoping not to be traced there, but somehow he was. I even suspect that the brilliant Lui was not what he pretended to be. Possibly, his task was to gain Mikie's trust and find an opportunity to look at his books. Or maybe it was just the opposite. Perhaps Lui was Mikie's partner in crime and together they developed some kind of large-scale scheme? Most likely, however, that Lui was in the business of a money laundering and attempted to use the Mikie's company in one of his operations, which accelerated Mikie's downfall. United Kingdom has a deregulated banking system and London is known as a money laundering capital of the world. If Lui was indeed a criminal, surely England was the best place for implementing his schemes. But what schemes? My guess would be no better than yours...

I still have a difficult time grasping one thing—why did Mikie do it? He was a bright man who loved his family (or at least so I believed) and enjoyed life. He could and did work very hard. He could have been quite successful just doing that. Instead, he lied to everyone, he misrepresented, he stole, and he was constantly fleeing from one country to another. He liked comfort in life, but he was always on the run. A strange man and a strange life... Where he is running to now? Is there any safe place left for him on this planet?

> Good judgment comes from experience,
> and experience comes from bad judgment.
> -Barry LePatner

I had moved to California and was living in an apartment in La Jolla, when early one Sunday morning the doorbell rang. It was an officer of the court who served me a complaint from a Texas court. The company Fresnel Devices of Fort Worth was suing Mikie and me for twenty thousand dollars to cover Mikie's unpaid bills as well as legal expenses. It turned out that Mikie had ordered the plastic lenses from this Texas company, and had not paid on time. When his business went belly up and was put in receivership, all his vendors lost money. Only Dick Klaytor, the owner of Fresnel Devices, had the idea of suing us both in the hope of collecting money from at least someone. I was told later that this Dick from Texas made a hobby out of suing people with or without reason, so long as there was some profit in it for him and his lawyer. Since Mikie disappeared without any trace, the Texan had decided to collect the money from me, not much caring whether I was in any way responsible for those bills.

I was very disappointed because someone had assumed that the crook Mikie and I were of the same kind. Besides, I was not in any way responsible for Mikie's bills and certainly had no wish to pay them. My participation in the dealings between Mikie and Dick had been limited to recommending Dick as a vendor and specifying what kind of lenses to produce.

Not knowing what to do about these legal papers, I asked several of my friends and neighbors for advice. One fellow suggested that I write a letter to the Texas judge explaining to him that I was an irrelevant person in this matter and should not even be considered a defendant in the lawsuit. This seemed to me to be a logical step, so I wrote a letter. Apparently, my mistake was that I was thinking in logical terms, whereas the legal system did not.

The judge ignored my letter, and a few months later I received notification of the hearing date. The same legally-knowledgeable friend of mine suggested that I should disregard it, arguing that, as I had never done any business in Texas and had no property there, I was not subject to the jurisdiction of a Texas judge. This also proved to be very wrong advice, because six months later the court marshal appeared on my doorstep again. This time, he served me papers from a California court. These documents I could not possibly ignore.

I went to a local lawyer and explained my problem to him. He told me that my situation looked very grim. It had nothing to do with the fact that I was not responsible for Mikie's actions and had not even signed any papers or promised to pay any of his bills. The lawyer explained to me that, in this country, the judicial system is concerned first and foremost with technicalities, justice being almost of lesser importance and secondary to the various formalities. Since I had not shown up at the hearing, he said, the judge had sentenced me by default to pay the whole amount that Mikie owed Dr. Klaytor, plus legal expenses. What was more, according to the U.S. Constitution, any judgment passed in one state was automatically recognized in another. This was called sister-state judgment.

"I suggest that you start collecting money and pay the plaintiff," said the lawyer. "The time for appeals expired long ago and the California court will not even bother hearing you. It will just accept the decision of the Texas judge and will automatically register it in the books of this state."

I refused to believe that the American justice could be so unjust, and decided to follow the old wisdom: never buy from a rich salesman and never hire a poor lawyer. So I went to the most reputable (read: expensive) law firm in San Diego and talked to lawyer effrey Blease. He confirmed the fact that my position was indeed none too bright, but suggested that if I paid his firm a retainer he would look into my case more carefully. "If all else fails," he said, "we can negotiate a settlement, so you may end up paying less."

Since the amount at stake was twenty thousand dollars, his initial retainer was five thousand. If, for example, the judgment against me had been one million, his retainer would probably have been around two hundred thousand. It seems the retainer has nothing to do with the time a lawyer will have to spend. He is just looking for a piece of the action. That is how lawyers work, I suppose. But what could I do about it?

"No way," I said. "Never in my life have I compromised over unjust things, and no matter how much it costs me, I will not negotiate or settle with this Dick from Texas. It's not a matter of accounting—it is a matter of morality and fairness."

The lawyer shrugged and said that it was entirely up to me.

I gave him a check for five thousand dollars and he set to work. There was no question about it, Jeffrey was a top-notch lawyer and knew what he was doing. He dug through the old case books and found that a judgment from one state was not always rubber-stamped by a sister state, as the textbooks claim. In one California case from 1924 a judge agreed to look into the matter of a case passed along from another state judge.

"This could be your lifesaver," said my lawyer. "If we convince the California judge to look into the merits of your case, rather than just rubber-stamp it, you might have a chance, because, after all, you are not guilty of anything."

He was right. On the day of the hearing, we arrived at the courtroom early in the morning. Our turn to face the judge did not come until that afternoon. The judge, having heard a few dozen cases by this point, was already tired and wanted to wrap things up quickly.

He said, "It's a sister-state judgment, let him pay."

And he was about to nail down his sentence against me with his wooden hammer, when my lawyer jumped to his feet and said, "But your honor, in 1924..."

Blah-blah-blah, he went, arguing that the judge must look into the merits of the case to see if the Texas judge was right. Our judge agreed and asked the plaintiff's lawyer to explain what this was all about. The California lawyer representing Dr. Klaytor told the judge that I had verbally guaranteed all Mikie's payments (which of course was not true). When he heard this, the judge said that California did not recognize verbal contracts over fifty dollars and dismissed the case. Indeed, the judge was so angry with the plaintiff for brining such a frivolous law suit that he even fined Dr. Klaytor four hundred dollars. That was that—and my only loss was my legal fees, which amounted to a very hefty sum on top of the retainer.

I learned a lesson from all this: dealing with the legal system is like playing a game—regardless of whether you right or wrong, you simply cannot win unless you play by their rules. And I learned another lesson as well from my sad adventure with Mikie-the-crook: be careful with whom you do business or eat sashimi. A person is judged by his entourage.

13

Instant Thermometer

While working for Timex Medical Products and seeing it fail, I tried to do whatever was in my power to rectify the problems. One of these was a series of drawbacks in the design of the medical thermometer which Timex had developed and planned to sell. Its prime technical feature was to be its speed. Timex wanted to make the fastest thermometer in the world, one which could measure a patient's temperature from the mouth in no more than ten seconds. The goal was a noble one, but the solutions were wrong. Technically speaking, everything was done correctly. However, that was not nearly enough. The project needed a revolutionary, rather than evolutionary, approach. What was in fact required was the reinvention of the entire method of measuring human temperature.

When the company realized that the thermometer was not good enough, the decision was made to dump it. Though not personally involved in the project, I was still interested in finding a possible way to bring it back to the drawing board. I went back to the basics I had learned in school years before. I knew that thermal energy can be moved in three ways: thermal conduction (which is used in conventional thermometers), convection (by moving warm or cold air or liquid), or thermal radiation (the same way the sun delivers energy to earth). Our bodies, like everything else in the universe, produce invisible infrared heat waves. These waves, if detected, can be used as a measure of the surface temperature of the object from which they emanate.

In the prototype of my Crib Death Alarm (apnea monitor), I had employed a special piezo-electric polymer film which produced electrical signals in response to mechanical stress. This wonderful material also had another property, a kind of side effect, which was totally useless for my project: it generated an electrical response when its surface absorbed heat. By putting together my basic knowledge of infrared heat transfer and my experience in working with this polymer film, I hit on the idea of devising an inexpensive medical infrared thermometer—that is, a thermometer which could measure body temperature even without contact, and could do so almost instantaneously. At the time I did not yet know quite how to go about it, but one of its features was very clear—it would be a damn fast thermometer, because it would measure heat waves which propagate with the speed of light.

I went to the vice president who was responsible for the Timex medical business and proposed that we look into infrared. The poor fellow was fed up with all these innovations and his job was already in jeopardy, so even thinking about anything exotic caused him almost physical pain. He looked straight into my eyes and said softly, "Jake, do me a favor, get out of here. I don't want to hear anything about any thermometer, or for that matter, any other great idea."

A few months later I left Timex and decided to take a chance on doing something on my own. In my "portfolio" I had several inventions which I believed could be developed into real products: a breast cancer detector, a wireless EKG electrode, an anesthesia monitor, a muscle properties detector, a personal monitor for ultra-violet radiation. And on top of all that I had my idea for developing a medical infrared thermometer.

When I was laid off, Timex paid me six weeks' severance pay which this time I used for a good cause. I did two things: raised money for a new business and built the first prototype of my infrared thermometer. This same duality marred my life for the next five years. I was continuously looking for funds

Instant Thermometer

and at the same time carrying on my research and engineering work. Unfortunately, my fund-raising efforts took up by far the greater part of my time, leaving very little for what I was good at—developing new products.

In my quest for investments, I made the acquaintance of several people whose business was deal-making. These wheeler-dealers gave me considerable input and a great many insights into various business arrangements of which I knew very little. Unfortunately, all these middlemen I met specialized in setting up connections between the U.S. and Far East-based companies. For this reason, I ended up with a number of none-too-useful leads in Hong-Kong, China, Singapore, and Japan, but very few in the U.S.

One such dealer, by the name of Earl Steiker, introduced me to the California venture capitalist Arthur Berg, whose specialty was medical instrumentation. I flew out to Los Angeles where Arthur received me very cordially. My wife Irena had never been to the West Coast before and so accompanied me on that trip. Mr. Berg and his wife invited us to their home, then took us out to an exclusive restaurant in Newport Beach and showed us several interesting places in Southern California. Mr. Berg and I had a very productive meeting at his office. Arthur told me that he was very interested in several of my inventions—except one. He said that my infrared thermometer did not fit into his plans, because he had just recently signed a deal to make another type of hospital thermometer which, he said, would make my infrared approach obsolete.

His thermometer resembled an EKG electrode which was to be attached beneath a patient's armpit at the time of his admission to the hospital. This disposable patch contained a small electronic circuit, but no batteries. To take the temperature, the nurse had to hold a special electronic box having a size of a small walkie-talkie close to the patient. The box transmitted a pilot radio-signal which would bounce back from the circuit beneath armpit at different frequencies, depending on the patient's temperature. Though I did not think the idea had any practical merit, I decided not to argue with Arthur, since I wanted him to invest in some of my other inventions which were of greater interest to him. After a very fruitful meeting, we shook hands and he said that he would send me his proposal within a week.

I flew back to Connecticut very excited and encouraged by Mr. Berg's business approach. Alas, a few weeks passed, then another, and another. I tried calling him, but his secretary kept telling me that Mr. Berg was either out of town, or in a meeting, or some other excuse. He never returned my calls. Thus did I encounter yet again that great deficiency of the American character—the inability to utter the simple word *no*.

In Connecticut I had a good friend, Leon Pintsov, who, after seeing my struggles to secure backing for my inventions, offered to introduce me to an acquaintance of his, Dick Gitlin, a partner in the large law firm of Hebb & Gitlin. Leon told me that not only was Dick well-off himself (have you ever seen a poor lawyer in America?), but his contacts in the business community could be a useful resource for me.

Initially, I was reluctant to meet Mr. Gitlin. Quite simply, I did not want to have any dealings whatsoever with a lawyer, for I have little regard for people of the legal profession. I see the great majority of them as hypocritical money-makers, rather than servants of justice. Lawyers serve first and foremost their own needs, and only then their clients. This does not, of course, mean that I fully sympathize with Shakespeare's appeal in *Henry VI*, "The first thing we do, let's kill all the lawyers." On the other hand, the simple fact that this sentiment was already prevalent four hundred years ago seems to say something.

From this prospective, Mr. Gitlin practiced the least honorable area of law. He was a bankruptcy lawyer—that is, a vulture who fed on dead bodies. Nevertheless, such is the situation in this litigation-happy country that a business cannot be born here without a lawyer, and it certainly cannot die without him. Despite my feelings, however, I was not in a position to be too choosy and decided to go and see Mr. Gitlin.

I met with Dick and told him my plans. After hearing me out, he offered to put down a small amount, but said that he would prefer it if I also talked to another lawyer, Alan Bartholdy, who was more knowledgeable in these sorts of things.

Alan was a lawyer only by education. Professionally, he ran the investment activities for Aetna Insurance Company. He was the perfect example of a venture capitalist—a man without the guts to get involved in any real venture. Around the time we met, he received an offer to invest in another invention and buy stock in a potentially profitable, yet risky, enterprise. He declined, and four years later, when that stock soared to some twenty times its original value, he almost killed himself in despair. Alan was a fine and trustworthy man, yet he felt an almost physical pain at the idea of anyone making more money then he (he was a lawyer, after all).

We met in Dick's spacious office on the top floor of what was then Hartford's tallest skyscraper. Following in Dick's footsteps, Alan said that he could also put down some of his own money, but could do nothing with respect to Aetna, as it would be ethically wrong to invest in the same deal personally and through his employer. He added that what I really needed was a partner or associate to take care of both raising money for the product development business and finding a company that might be interested in obtaining a license for at least one of my inventions. I could not agree more. He said that he knew of another lawyer (what a chain reaction!) from New Jersey. His name was Ralph Lilore and he was the business partner of a young man (finally, not a lawyer!). They had established their partnership for the purpose of raising funds for businesses developing new products, bringing the stock value up, and then selling those businesses at large profits. Not a bad idea for making money and just what I was looking for.

I asked Alan if he knew of any previous successes by these two partners, and he told me that so far there had been none, as they were just getting started. Of course, I would have preferred dealing with someone with a proven track record, but after giving it some thought, I decided to give them a chance.

> And be these juggling fiends no more believ'd,
> That palter with us in a double sense;
> That keep the word of promise to our ear,
> And break it to our hope.
> -*Shakespeare,* Macbeth.

Meanwhile, exploring another avenue, Alan set up an appointment for me with Mr. A. Pelligrino, the chairman of a medical instrumentation company. This Mr. Pelligrino was interested in hearing about my anesthesia monitor. The company was located in the town of Wallingford, only fifteen minutes from my home. When I arrived there, I was greeted very cordially. My hosts—the chairman, the president, and the director of engineering—gave me a tour of the company, after which we sat down to a lengthy meeting. I presented my proposal and said that I was interested in licensing the monitor to their company in exchange for future royalties. In addition, I offered my services on a contractual basis to help them in the development process.

They then made me a counter-proposal. They offered me a job. The position was director of research and development, and was contingent on the rights to my monitor being transferred to them for nothing. This, of course, was not what my partners and I were after. Mr. Pelligrino said that he understood my position, and since they really wanted to work with me, he asked me to wait a couple of days until they were in a position to come up with the most appropriate formula for our relationship. The chairman promised to call me on Friday. Pleased by this dynamic approach, I said sure, I could wait till Friday.

Sounds good, doesn't it? That conversation took place in the fall of 1983, and I stopped waiting for his call many years ago. Just as with Arthur Berg, all my attempts to get some kind of response from Mr. Pelligrino or his subordinates yielded nothing. They simply ceased all communication with me. Once again, it was the old problem—the verbal inability of an American businessman to say the word *no*. I can understand that they might not have liked my proposal or that they might not have liked me, but why not say so? The pattern had become so annoying that I decided to stop putting much hope in such meetings. It was time to get partners with thicker skin than mine. Maybe they would be more tolerant of that brand of "politeness" or know how to handle it better. In any event, I could not deal with it anymore, so I called Alan and asked him to bring in Ralph Lilore and his young partner from Cleveland.

> Never mistake motion for action
> -Ernest Hemingway

As on previous occasions, we met in Dick's law office. This time, besides him, Alan and me, we had two out-of-state visitors: a stocky man of about fifty—Ralph from New Jersey—and a young fellow with straw-colored hair—Steven Jonson from Cleveland. Steven was only twenty-five or so and could hardly keep his mouth shut. His manner of speech reminded me of a machine-gun rattling off some thousand rounds a minute. Fortunately, he was not only a fast-talker, but a fast-thinker as well. Otherwise, it would have been impossible to communicate with him.

During the meeting the decision was made to start a new corporation to promote my inventions. I was to serve as its president while the rest of the party agreed to become members of the board and the initial shareholders. For fifty percent of the stock, I contributed several of my inventions and a set of laboratory test equipment which I had purchased over a few years for use at my home. The other founding members contributed $5,000 each for the remaining half of the new company's stock. We decided to name the new business Fra-Med, as a contraction of my last name and the word "medical".

On second thought, I changed one letter and it became FreMed, Inc., because the word *framed* sounded too suspicious.

Later, I realized that I should have been less generous in giving away so many shares in the new company to persons who had thus far promised a great deal of help but contributed very little. They may have had the best intentions in the world, but they were simply not able to deliver soon enough or as much as one might expect. Even if one's partners put up some money, this is no reason for giving away the store. All promises must be valued against performance, and all stock should be vested upon demonstrated results. Had I done this, not only would it have protected my part of the venture, but it would have given them strong incentives to work harder. It took me a good many years to learn this lesson:

> **Do not share stock with your partners until they prove their efficiency**

I rented a large room in an old brick building. It was located in the heart of the New Haven slums. The city business development council had surrounded the entire area with a tall fence and renovated a few nineteenth-century buildings to house several start-up companies operating on shoestring budgets. The place was managed by the New Haven Science Park Corporation, a kind of incubator for entrepreneurs. Most convenient for me, Science Park was associated with Yale University, and I automatically was eligible to obtain medical insurance from the university and use the university libraries and certain other technical facilities.

At a second-hand store, I bought some furniture for the laboratory—a couple of desks, several tables, chairs, and file cabinets, all of which had seen better days—and began working on my projects. I needed a technician and an engineer, so I placed a few ads in the local newspapers, and within a couple of weeks I had hired two young men: Mac, an electronic technician, and Ernie, an engineer. To pay their wages, medical insurance and various taxes, we needed more money. My new partner Steven quickly went out and sold more FreMed stocks

Instant Thermometer

to several of his friends and relatives. Obviously, our shares in the company were thus being diluted still further, but what could I do?

> Those who dare, do;
> Those who dare not, do not.

At first, we did not know what invention we might be able to find a buyer for, and so decided to make demo prototypes for at least a few of them. This was an enormous task for just three people. Within a few months we had built a wireless EKG electrode and also a test instrument for evaluating several ideas behind the anesthesia monitor. In addition, we put together the first "blackbody." This device got its name from its appearance—if one looked straight into it, it was indeed damn black. This particular piece of equipment was an essential step in the process of developing an infrared thermometer. Since the thermometer was supposed to measure the temperature of a surface without touching it, we had to design a test surface whose temperature we could measure precisely. Without such a machine, it would be impossible to evaluate how good or bad our thermometer was. And we had to design it ourselves for the simple reason—no one had ever attempted to devise a medical noncontact thermometer before, so there was nowhere to buy specialized test equipment for it.

We made the very first blackbody out of a stainless steel dish containing stirred machine oil. Into this we placed a mercury thermometer and an electric heating element attached to a thermostat control. The heating element warmed the oil, the oil warmed the dish and the surface of the dish could then be used as a "blackbody." Actually, this was not the best way to go about it, and after studying several books on physics, I designed another version, a so-called cavity blackbody. We constructed it from a piece of a steel pipe wrapped inside a layer of heating wire and some thermal insulation. This new blackbody worked reasonably well, but after only a couple of days, something went wrong with the control circuit and the insulator caught fire. The entire laboratory was filled with black smoke. Fortunately, nobody from Science Park noticed it, otherwise they might have kicked us out.

The anesthesia monitor had to take its data from the tips of a patient's fingers. We designed a series of special electrodes and electronic interface boxes. I planned to send all the equipment to my friend Professor Simon Gelman, who by this time had been appointed chairman of the anesthesiology department at the University of Alabama in Birmingham. It was Simon who had first suggested that I develop such a monitor, and he had agreed to run the first clinical tests. I did a great deal of mathematical modeling and was fairly confident that the device would work. In order to bounce some ideas around with people working on similar types of monitors, I traveled to Yugoslavia where, in the ancient town of Zadar, an international conference on related subjects was being held. There, I had an opportunity to see what people in Japan, Austria, the United Kingdom and many other countries were doing in terms of anesthesia monitoring.

Needless to say, it was an enormous project requiring much effort and time. Also, it drained our precious resources. The longer I worked on it, the clearer it became that there could be no successful completion of the project unless I was associated with a medical research facility—not on an intermittent basis, as I was planning to do with Simon, but essentially full-time. This type of monitor simply could not be designed and tested in a laboratory—it had to evolve from a continuous process of R&D at an established medical facility.

Also, there was one other issue that I could not ignore.

The development of medical equipment for use in emergency and operating rooms, ICUs and other medical facilities is a long, highly risky, and very expensive enterprise. The patient's well-being and even life often depends on such equipment, and so you want to be damned sure that it is going to function almost flawlessly under virtually any circumstances. Even if you are confident that it works, the Food and Drug Administration will demand that you convince nearly everyone else as well. This could take years and years of work. Who were we going to find to finance such an enterprise? The best source of money would be grants from the National Science Foundation, the Small Business Administration, or the National Institute of Health. I applied to all three agencies, but was rejected. Quite simply, they felt that such a project was too complex for a start-up company and that no grant would be sufficient to keep the business afloat for several years without additional infusions of capital.

Considering all these circumstances, I came to the conclusion that the project had to be abandoned. I simply could not afford to do so many things at once.

Being a risk-taker, an entrepreneur must first of all be a risk-avoider. That was how I learned an important lesson:

> Do not work on several projects at once.
> Focus on the most promising and least risky.

Having understood this, I stuffed all the prototypes of my anesthesia monitor and wireless EKG electrode into a big cardboard box, labeled it "Museum," and put it away on a shelf. From that moment on, my one goal became the development of a medical infrared thermometer. Steven, Alan, and Ralph all agreed with me that this was indeed the invention that carried the least financial risk and offered the greatest potential financial reward. And isn't that what any investor likes to see?

The first prototype of an infrared thermometer I put together even before FreMed was started. It was a very simple device: its housing I made from a butchered audio tape cassette, the sensor was fabricated from a piece of a special polymer film clamped between two copper wires which, in turn, were attached to an electronic circuit and a digital voltmeter. I realized that the film responded only to changing infrared light, so to make a change, I built a sliding door out of a piece of painted Plexiglas and used a rubber band as a spring to pull it back. That was my first working model, which I wanted to test and further develop in designing a medical oral infrared thermometer. Yes, I still thought that the only place for accurate temperature measurement was the mouth. But how to go about this with my prototype? I selected the tongue as the place for temperature taking.

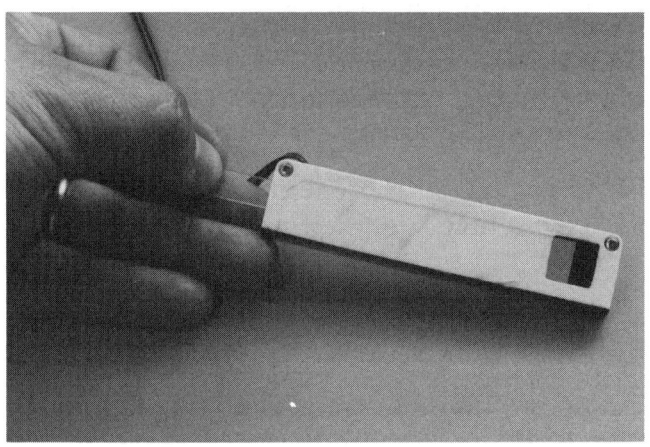

The first prototype of Instant Thermometer

It went like this: before taking a measurement, I pulled out the sliding door on my thermometer (we called that door a shutter), opened my mouth, placed the square window of the thermometer on the surface of my tongue, and released the shutter. The rubber band pulled it back, the door closed quickly, the polymer film generated an electric spike which was measured by the electronic circuit, and the voltmeter showed the numbers graduated in degrees Celsius.

The thermometer worked very fast, within a fraction of a second. Thinking about a name for this new product, I decided to call it *The Instant Thermometer*, a name which stuck and is used to this day. It seems as though it has even became a generic term, and if so, it will be my contribution to the English language.

To my joy and pride, the prototype worked pretty well and I was able to measure temperature with it. However, my excitement dissipated all too soon and I felt disaster looming. Simply measuring temperature was not nearly enough. The most critical requirement was not just to make a thermometer, even an accurate one, but to make a *practical* thermometer.

First of all, sticking one's tongue out was not practical from any point of view: most sick people, children, the elderly, and many others simply couldn't do that. Besides, even if they did, the thermometer had to be placed on the tongue almost instantly, within a half a second, otherwise evaporation from the wet surface would lower the temperature of the tongue and my thermometer would read it as three or four degrees lower. This was a totally unacceptable design and I had no desire to show it to anyone. Demonstrating such a prototype would surely be a suicidal move. I felt discouraged and did not know what to do. I had found a way to design a fast and inexpensive medical thermometer, but my design was nothing more than a totally useless gadget.

I was worried and confused. I had made a promise to my investors and they had trusted me, but now I had no idea how to proceed further. Some people, when they find themselves under pressure, lose their abilities and give up. Fortunately, when I am cornered, I become more efficient and even aggressive. I attacked the problem from all angles. I thought about it day and night. The days went by and I still had no answer. I became quite irritable, and once again my wife suggested that I stop killing myself and find a quiet and secure job, like all "normal" people do. In my dreams, I saw invisible infrared rays shooting in all directions. I tried many other versions of the design. Naturally, in the laboratory I was my own guinea pig. With the various prototypes we built, I measured temperature in all parts of my body, hoping that some other site could be used instead of the tongue. I considered taking a temperature at the neck, the nasal cavity, the eyeball, the temple (which would be okay only for bold people). Some of these places correlated with an oral temperature better than others, but still the difference was quite significant and, what was even more important, skin temperature was strongly influenced by ambient temperature and that was no good for a practical device.

Second Prototype of Instant Thermometer

One day I thought about the ear canal. The ear has several interesting features, being a dry cavity very little influenced by ambient air. Also, its shape is similar to a blackbody, that is, it is a cavity—a nearly ideal place for infrared temperature measurement. I looked through anatomy books and found that the ear drum shares a blood supply with the brain. I knew that when doctors measure temperature, they really are not interested in learning the temperature inside the mouth. The brain—that is the goal of the measurement. Hence, the ear was an attractive choice: it was close to the brain, easily accessible, clean (in most people), and had a low risk of transmitting infection. The clairvoyant prince Hamlet had a vision that his uncle Claudius had poured poison into his father's ear and the king died instantly. I suspect that either Hamlet was paranoid (which is what everybody thought), or else in Shakespearean times people possessed a different anatomy, because contaminating the body through the ear is much less likely than through mouth. So, I decided to use the ear.

However, I had no way of measuring temperature in the ear canal: my prototype with its sensor and shutter simply would not fit in an ear. The thermometer might well work in an elephant's ear, but I wanted to use it on people. Somehow, I had to take infrared light from the ear canal, channel it outside, and couple to the sensor. In the end, the solution I found was very simple (everything good has to be simple). I decided to use a reflective tube which would work as a pipe for the infrared light. This thin one-inch long tube would be inserted into the ear canal. Its outer end was positioned next to the shutter and sensor. The inside of the tube had to be very shiny, like a mirror. When you placed it in the ear canal, infrared light would pass into the open end, bounce back and forth off these shiny walls in a zigzag manner and exit right to the sensor of the thermometer.

In order to test this idea, I took a piece of aluminum foil and wrapped it around a pencil to form a metal tube. Using a few drops of Elmer's glue, I secured it to my prototype, and once the glue dried, I calibrated the sensor in a blackbody and inserted the tube into my ear. The effect was excellent. The meter gauge showed a temperature of near 98°F, which was my normal body temperature. At that point, I was convinced that, from now on, temperature could and should be measured from the ear canal.

I built a few more models, did a great deal of thinking and calculating, and in September of 1984 I felt that I knew exactly how the ear thermometer had to be designed and built. I hired a patent lawyer and we began working on the patent specifications.

Meanwhile, Steven did not waste any time. Very energetically, he gathered more investments from his friends and acquaintances, and especially from his father Lindsay Jonson. Without that money, FreMed, Inc. would not have stayed in business even a few months.

We lived and worked on a shoestring budget, counting every penny and constantly looking for less expensive ways to obtain desired results. However, the funding we received from selling equity was not nearly enough for full-scale research and development. Yes, we found many less expensive ways to do our work, but these ways were often not the fastest or most efficient. Often we had to spend months on something that might have been accomplished in just weeks, if we had simply had more money.

My primary activity was not research and development, as one might have expected. A great deal of my time I had to spend writing descriptions of the invention, drawing pictures for presentations, preparing business plans, writing proposals, doing market research, and traveling all around the country meeting people.

> **Inventing is 5% creativity and 95% promotion**

Steven and Ralph did their own selling and kept me awfully busy. Primarily, in order to raise money, they targeted the venture capital groups. During the first two years of FreMed I had nearly a hundred meetings with this type of investor. They ranged from rich individuals to such reputable venture capital firms as Syrinx from the United Kingdom and Canadian Enterprise from Toronto to the Ventec, Avalon, Sprout, Clarion Capital, and Euclid Partners groups in the U.S. The difficulty was that I had to be prepared for each meeting and to tailor my presentation and business plan to their individual interests.

Dealing with these "venture capitalists" was, while socially pleasant (after all, they were all nice people), a totally useless waste of time and money. A venture capital firm is not the right source for money to finance an innovative enterprise. All these endless meetings (sometimes, I had to travel to New York City several times a week) all too quickly drained our more than modest funds and kept me from focusing on my research work. From that experience I learned another lesson:

> **Don't deal with venture capital companies to raise money for a start-up business**

Besides the venture capitalists, we contacted a number of manufacturing firms which Steven and Ralph tried to persuade to enter into a licensing agreement with us. Depending on the type of business, we offered them different inventions (from my "Museum" cardboard box). Most of the medical equipment companies were interested in the wireless EKG electrode or in the anesthesia monitor, but somehow we were never able to progress beyond two or three meetings. After that, our prospective partners would invariably lose interest. I could not understand what the problem was. Why would people show a great deal of interest at first, and then shy away? I remained puzzled, until one man explained to me what was at least his personal reason.

Steven had arranged for us to meet a private investor in Stamford, Connecticut. His name was Jack Whitehead. At our first meeting, I made my technical presentation and outlined the results of my marketing study. Steve then discussed the financial and investment aspects of the deal. Jack said that he liked what he heard, but needed time to think it over. We decided to meet again four weeks later.

When Steven, Ralph, and I arrived for our second meeting, Jack Whitehead very quickly made it clear to us that instead of the $200,000 investment we had talked about before, he would consider putting down no more than $15,000. This was not nearly enough and, very disappointed, we prepared to leave. As I was going out, Jack stopped me in the lobby and said that he had changed his mind after the first meeting because he did not feel comfortable with our group. I asked him to be more specific.

"You see," he said, "I believe in your technology. Also, I clearly see that the market for at least two of your inventions is there. I even think that your demands are modest and reasonable. What I don't like, is the way your young friend [he meant Steven] talks. He talks faster than I can think. I have an uncomfortable feeling that his manner of talking is specially designed to steer me away from my own thoughts. He is bombarding me with words. My gut feeling tells me that he wants to hide something from us and I don't like that. That's why I decided to pass."

It had never occurred to me that Steven could make such a wrong impression. Still, I could not ignore Jack's words and decided to mention it to Ralph. He agreed with me that we should take these comments very seriously. From that time on, whenever we met with other companies or investors, Ralph assumed a more active role and Steven a more passive one. Instead of an asset he had become a liability.

> In a revenue-raising effort the Emperor Vespasian instituted a tax on public urinals. Among many opponents was his own son Titus. He bitterly commented that the new tax stank. Vespasian, a man with a great sense of humor, held some money from the first payment up to his son's nose and said, "Smell it! Money doesn't stink."

During the years we spent searching for funds, I was a frequent visitor on Wall Street. I met quite a few strange people there: very smart, very quick, very rich, and very malicious. This was the era of the yuppies making their quick bucks. One man, by the name of Peter Katevatis, was involved in the manipulation of public medical companies. He made me a proposal for the merger of FreMed with a company he owned called General Cardiology. When I asked him what this company did and whether I could visit it to see if his proposal made sense from a business standpoint, he said no, the company existed only on paper. Its sole asset was a five-million dollar loss. In effect, Peter was selling a loss, which could be used as a write-off for tax purposes. For this "treasure" he wanted 70% of the combined business.

Another crook, Harris Freek, was introduced to me by Harry Barres, an executive at Drixel Burnham Lambert. Harris had an investment firm, Bridge Capital, which primarily traded in junk bonds. Together with Drixel, he devised a scheme which, for reasons I cannot fathom, he believed anyone would be happy to leap on.

The idea was simple. Like Katevatis, Drixel Burnham Lambert and Bridge Capital wanted to merge FreMed with one of their paper companies that traded publicly for a penny a share. Since I owned an attractive technology whose benefits could be grasped very easily by the general public, they planned shortly after the merger to arrange a good deal of publicity for my invention. Presumably, the one-cent stocks would then skyrocket, at least in the short run, but that would be enough time for the controlling parties to dump their holdings, skim the cream and get out.

Harris was very unpleasant man. He was relatively young, no more than forty, with a bony posture, a terribly pale, almost corpse-like face, and the slimy personality of a lawyer. He spoke briskly but quietly, without any emotion. His eyes were constantly shifting off to the corner of the room, never looking at me.

When he finished describing his plan, I thought that I missed something.

"Harris, wait a minute, but what about the thermometer, what about the business?"

He must have thought he was dealing with an idiot—at least that's how he looked at me, for the first and the last time.

"We are not talking here about any busi-i-iness. We are talking about making mo-o-oney! Are you interested?"

I shook my head. Harris immediately stood up and, without saying good-by, walked out of the room, no doubt to chase another, more practical fellow. I have not seen him since. It gave me an almost physical pleasure when, a couple of years later, the rich and powerful Drixel Burnham Lambert was shut down by the FCC for illegal business practices.

Along with our fund-raising efforts, we were also searching for a company that might be interested in licensing one of my patents. The idea was that such a company would license the patent, while FreMed would be subcontracted to carry out the research and development.

One day, Steven told me that Jim Loveland, a vice president with an Ohio company called Mini-Q, was going to visit my lab in New Haven. His company produced medical monitoring equipment and was looking for new ideas to expand its product line. He was primarily interested in the instant thermometer, and before we proceeded further, he wanted to see for himself what the prototype looked like on the lab bench. At that time, I was working on a second prototype, with a better optical and mechanical design. It was to be a hand-held thermometer which could be used for taking temperature from a human ear. Unfortunately, the new prototype was not ready by the time our visitor from Mini-Q arrived.

Mr. Loveland spent a few hours with me. I showed him my old prototype and explained the course of our work and what had been accomplished so far and what still remained to be done. He left, and a couple of days later called Steven and suggested that we all come to Ohio to present our product to the chairman and the president of the company. He said that the interest was high and that they even wanted to begin the preliminary negotiations.

I flew to Cincinnati, where I met Ralph who had come out from New Jersey. We rented a car and at the end of a two-hour drive found ourselves warmly greeted by the top management of the company. By this time, I already had my second prototype up and running and it worked very well. It was still a far cry from a production model, but nevertheless it was a perfect demo unit for selling the idea. I proudly demonstrated it to our hosts.

The meeting was productive and we discussed a number of issues and even outlined a future business plan. I flew back to New Haven feeling that our long struggle might be coming to an end and that the business might soon be propelled to a new level.

However (as had happened so many times before!), silence fell. None of us could get any further word from any of the top managers of the Mini-Q company. Finally, Steven managed to catch Mr. Loveland on the telephone. The response he got was astonishing:

"Our company has enjoyed a high reputation for a number of years and we don't feel that it is appropriate for us to deal with a bunch of con-artists like you. When, just two weeks ago, I visited New Haven, the only thing I was shown was a cumbersome primitive prototype which could take only very inaccurate temperatures from a blackbody. But now, you come to us with a beautiful and accurate thermometer which was a dramatic leap from what I had seen just a couple of weeks before. It is impossible to make that much progress in such a short time. It's obvious for us that you bought some expensive industrial infrared thermometer, took the guts out of it and repackaged them into your home-made case. You did this on purpose to fool us, but we were smart enough not to believe your scheme. Good-by. Don't call us."

Needless to say, we were shocked, awfully offended, and angry. Unfortunately, the only revenge we could contemplate was to succeed and leave those "smart" people on the losing side.

I made presentations to so many companies, that all those meetings became for me like the stage performances of an actor who has played the same role every night for many years (my acting training came in quite handy). Steven or Ralph or both of them accompanied me on nearly half of these meetings. We visited Chesebrough-Pond's, Duracell, Tenax, Capintec, Medex, Beckton Dickinson, and Clairol—just to name a few.

Dr. Stewart Kurtz, vice president of technology at Clairol, told me that he had recently seen a prototype of another medical infrared thermometer which was being developed in California by a company with the modest name of Intelligent Medical Systems (a funny name indeed, as though other medical systems are not intelligent). This was how I first heard of a competitive group. That company, known by its abbreviation IMS, worked practically in parallel with me but apparently had better luck in raising money, and its ear thermometer was put on the market a few years earlier. Alas, our approaches were totally different. IMS developed a sophisticated and indeed intelligent ear thermometer for hospital use. It was a beautifully and professionally designed medical instrument, though too big, heavy, and very expensive. By contrast, my goal from the very beginning was to make an infrared thermometer for consumers. That is, it had to be accurate, but at the same time affordable for the general public, not just for rich hospitals. After all, a thermometer is the most essential diagnostic tool for self-care.

In the 80s, there was a popular TV commercial in which a middle-aged man with neatly groomed face stood in a robe in front of a bathroom mirror. Holding an electric shaver in his hand and looking straight into the camera, he said that after buying the shaver, he liked it so much that he had also bought the company that made that shaver. The name of the man was Victor Kiam and the company was Remington.

One evening I was browsing the shelves of a local bookstore when I noticed the familiar neatly groomed face on the cover of a book entitled *Going for it! How to Succeed as an Entrepreneur*. Naturally, I bought the book and read it from cover to cover. One of the stories Mr. Kiam told was that of a prolific inventor who came to Playtex, where Kiam was working the time, with one of his inventions: two pieces of fabric that stuck when pressed together and then unzipped when pulled apart. Neither Playtex nor Mr. Kiam himself went for the invention, which became a new type of zipper known as Velcro, an extremely useful and popular product. In the end, Victor very much regretted that he had missed that opportunity and stated that "...I didn't let many like that slip by me again."

I thought to myself, "Do I have a product for Mr. Kiam! A man like that won't pass up the opportunity to revolutionize the way people measure their temperature." I sent him a letter suggesting that we meet and talk about my invention. A few days later I received a telephone call from Bob Durkin, Kiam's son-in-law, who also served as a product director for Remington. Bob asked me to come over and talk with him and Victor.

Remington was located in Bridgeport, Connecticut, a short drive from New Haven. Victor looked very much like the man I had seen on the book jacket and on TV. There was even the same bathrobe featured in his commercial hanging in his office. Apparently, his personalized TV ads were part of his nature. He liked big shows and enjoyed mingling with top level politicians. We met quite a few times, and every time he tried to do something impressive, like putting then Vice President George Bush on speaker phone to chat when he called during one of our meetings.

Victor liked the idea of the instant thermometer, but apparently had no strong gut feeling for "going for it."

He said, "Look, in reality the business that you propose consists of two businesses: one is the electronic infrared thermometer that I probably can make and sell, the other is a disposable probe cover. The cover is a small plastic part and I don't do plastics myself. All my disposable plastic shavers for the hospital market are produced by the 3M company of Minnesota. Let's do this. I'll call 3M and if they decide to take over the disposable part of the business, I'll take over the thermometer. Is it a deal?"

Victor even sent me a draft of the proposed agreement and did what he promised. He contacted 3M, and 3M's man in charge of new business development came to New Haven and met with me. About a month after his visit, he called me back and said, "It's a policy of 3M that we don't get involved in any business that can't generate at least three million dollars a year in sales. We've estimated that the total size of the U.S. market for your probe covers is about three million dollars. It's too small a potential for us and we've decided to pass."[1]

---

[1] Thermoscan, the company that now makes my instant thermometer, sold near eight million dollars' worth of disposable probe covers, a highly profitable product, in 1994, and that is just a small fraction of the total opportunity.

This response automatically triggered the dropping out of Victor Kiam. That lovable man certainly had a great talent for missing opportunities.

Steven set up a meeting with Norelco, a company which competed with Victor Kiam in the sale of electric shavers. We made the usual presentation, then another one, and a few more after that. Norelco's interest grew very high. Its engineers evaluated our technology, its patent councils analyzed the strength of the patent, the marketing department concluded that a large market could be created for the device. This process took several months of intensive work on both sides.

Soon, FreMed ran out of money and I could no longer pay either myself or my employees who had continued to work on the project while I was wasting my time on endless meetings with various representatives of Norelco. Steven, in turn, had run out of friends and relatives from whom to raise more money. I became quite frustrated with his efforts. It turned out that this fast-talker, whose job was to provide the financing for our business, could only raise money from his own father. And even that source became less and less available. Lindsay Jonson had already invested over one hundred thousand in FreMed, after which he had decided to sit and wait. Fortunately, he was a man who could not stand to see anyone cross his path. Whenever I made some promising contact with a reputable investor, he sent me another check, just to retain more control (in exchange for more FreMed stocks, of course). This was probably the only useful outcome of my meetings with Victor Kiam, since whenever Victor interest went up, Jonson put more money down.

When Norelco concluded that it made sense for it to enter the thermometry market, our lawyer and the Philips legal department (Norelco was owned by the Dutch company Philips) began working on a draft of the contract. It took an enormous amount of time, almost a year, to finalize the deal and to smooth out all the minute disagreements. To stay afloat, I was able to secure a loan from the city of New Haven, whose business development council was very supportive of a small struggling company. This money kept us from bankruptcy, but it was not nearly enough to continue any serious development of the instant thermometer.

Finally, at the end of May of 1986 I received an overnight packet from Norelco. It contained the final version of the contract, which I had to sign and send back to the president of Norelco for his signature. After that, we were supposed to receive a check and Norelco would start things rolling. I went over the contract, signed it and mailed it back to Norelco. On the morning of May 31, in great anticipation, I waited for a telephone call from Bob Gaines, Norelco's director of marketing, who was to call me as soon the contract was signed.

Bob called me around noon. "I am sorry," he said. "I must apologize. At the very last minute our president made the decision not to sign the contract."

It was a great blow for all of us. The excuse I was given was that the decision had been made in Holland and the president was just following the orders of his overseas superiors. I believe, however, that there were more subtle reasons at work. Apparently, Norelco had some very serious internal problems, because a month later I learned that its president, who had refused to sign our agreement, was forced to take early retirement and one of the top financial people of the company committed suicide. Shortly after that, Bob Gaines called me again and invited me to come back to Norelco. I was fed up with them and said no.

But in June FreMed was technically bankrupt, our debt was almost a quarter of a million dollars, and we had very few tangible assets. I had no choice but to lay off our employees, including myself. Since FreMed was a legitimate business which paid unemployment tax, Ernie and Mac were able to collect unemployment compensation. I, however, got yet another shock when my application was turned down and I was told I would receive no unemployment payments. The reason I was given was in the brilliant bureaucratic tradition: "We don't pay unemployment to people who quit their jobs. You were the president of the company and you laid yourself off, that is, technically you quit your job." My argument that in that case the state should refund the unemployment tax I had paid for myself, fell on deaf ears.

Difficult times arrived for my family. My wife made some money by playing for the New Haven Symphony, but that was not nearly enough to pay the mortgage and our other expenses. Nevertheless, I was still hopeful. I knew that only for the dead does life never change. What had happened to my enterprise I saw as a sad but temporary event. I was poor by accident, so to speak, not by design. Life went on, and if you have the will, you can always pull yourself out of a hole. I liked the remark of Mike Todd, the theatrical entrepreneur, "I've never been poor, only broke. Being poor is a frame of mind. Being broke is only a temporary situation."

I considered various scenarios for getting out of that difficult temporary situation. One thing, however, I did not consider—quitting.

> Keep on going and chances are you will stumble on something...
> I have never heard of anyone stumbling on something sitting down.
> -Charles F Kettering, American Inventor

With Norelco we lost fifteen months and about a hundred and fifty thousand dollars. Even more disappointing was the fact that, with a deal supposedly pending, we had stopped looking for another licensor. We found ourselves in a much worse position than when we had just started out two years before. Time works against inventions—they grow older much faster than people.

I decided to wait for a couple of months, and if we did not find any other source of money, or some other company willing to license the patent, I would have no choice but to dissolve the business and take a regular job to support my family. The two fellows who worked for me also decided to stay. Of course, they continued to look for jobs, but they still came to the laboratory almost every day and worked without pay.

Instant Thermometer

One hot summer day, I got a call from Science Park, my landlord (who, by the way, had generously allowed us to remain there even after we had stopped paying the rent and utilities). The caller asked if I would mind meeting with a correspondent from *Business Week* magazine. I said sure, why not? The correspondent was a young man who had heard about the deal with Norelco falling through and wanted to write a story about me. We set up an appointment and he stopped by together with a photographer. I put on my bow-tie in order to look a little bit the mad scientist, and they took a series of pictures.

SOVIET EMIGRE FRADEN: WITH NO PRODUCER FOR HIS THERMOMETER, HE'S JUST HANGING ON

Photograph from Business Week Magazine

The article about the instant thermometer—complete with a picture of me sticking it into my ear—first appeared in the international edition of *Business Week* and then a few weeks later in the domestic edition. Almost immediately, I began receiving telephone calls from all around the world. Most of the callers were interested in investing, however, the majority of them were from Saudi Arabia (no doubt they were looking for a good place to park their petrodollars). When the American edition appeared, the first call I got was from my competitor-to-be, the IMS company of California. They offered to buy the patent from us—for ten thousand dollars. Fortunately, I was not that hungry.

The next call was from Dave Smith, a manager at Teledyne Water Pik in Fort Collins, Colorado. He said that they wanted to talk to me and asked if I could meet with them the very next day in my office. I told him to come on by, I would be happy meeting with him. Oh boy, what a little good publicity can do for you! Steuart Britt once wrote, "Doing business without advertising is like winking at a girl in the dark. You know what you are doing, but nobody else does." Thanks to *Business Week*, I winked at the whole world.

I called Steve to tell him that I was expecting visitors from Colorado the next day and that perhaps we might be able to rectify our hopeless situation.

About ten minutes later, Steve's father called.

"Listen," he said, "don't make any promises to those visitors. I might consider taking over the entire deal."

This was an interesting twist, though very much in line with Mr. Jonson's character. As he saw it, this sudden interest in the instant thermometer was a sign of a great opportunity. He had already invested in my business, but not nearly enough to gain total control, so he was hinting at me not to rush off with any other suitors.

The very next morning, Dave Smith and the vice president of marketing for Water Pik showed up. I demonstrated the technology for them and said that we were merely looking for someone to replace Norelco. The visitors took their notes and left.

Two days later, Dave called from Colorado and asked me to come out there for a meeting with the company's top executives. He said that their interest was very high and that they wanted to negotiate a deal with FreMed right away. He also said that Water Pik had already booked me on a flight for the following morning. That was darn fast, much faster than anything I had ever seen. I called Steven and told him that I was going to Fort Collins.

And again, ten minutes later Mr. Jonson called me (evidently, Steven had a good rapport with his dad).

Lindsay said, "Jacob, don't go to Colorado. I'm telling you, I may consider taking the whole deal myself."

"Lindsay," I said, "I appreciate it, but it's still worth going to Water Pik and hearing them out. I assure you that I am not going to sign anything, just listen. When I come back, you make me an offer and we'll decide together what to do."

He, however, was listening only to himself. "Did you hear what I said? Don't go to Colorado."

The next morning I drove to LaGuardia airport and boarded the Continental flight to Denver. I thought about Jonson's words, and could not understand why he had insisted so strongly that I should not go. I supposed he felt that a good deal might be slipping through his fingers. If he wanted to make me an offer, why not just make it? Instead, he was pressuring me to decline an opportunity, something that was strictly against my nature—never ignore any opportunity.

The flight attendants had prepared for take-off, the pilot started the engines and we were about to begin taxing to the runway, when the engines were killed and the stewardess announced, "Is Mr. Fraden among the passengers? Sir, you have an urgent telephone call."

What the heck was this? I was puzzled. Who had the power to stop the plane, and what kind of emergency was it? At that time, I did some consulting for the CIA, and it occurred to me that I was probably needed by them for something really important. The stewardess said that I could take the call at the gate counter and that the plane would wait for me.

It was Lindsay Jonson.

"I am telling you for the very last time, get off that plane and don't go to Colorado!"

"Look, Lindsay," I said, "what's all this fuss about? I promise you, I'm not going to sign any paper, I'm only going to listen."

"If you don't come home immediately, if you dare go to Water Pik, I will never speak to you again."

That was the worst thing he could say to me. Whenever I am intimidated, I tend to strike first. From that moment on, it was no use continuing, so I said, "It's up to you, though I am going."

Lindsay was born to big money and thought that he always held his fortune by the tail. His nature was geared for commanding, not convincing. If he had something to offer, he ought to have appealed to my common sense and I would certainly have listened. Voltaire once said that common sense is not so common, but that, at least, is one quality I have always possessed, and Jonson knew it. But he had decided to treat me as a general would treat a private, forgetting that I was not his private. Unfortunately, his rudeness and my obstinacy resulted in a diversion of our enterprise and the loss of several years.

The meeting in Fort Collins took the entire day. The executive vice president, Arnie Huppert, proposed licensing the patent on my Infrared Thermometer to Water Pik in exchange for a down payment and royalties. The company wanted to use its internal resources to develop and sell the infrared thermometer in the consumer market. We negotiated every last detail of the deal and agreed that a final contract would be drafted within a few weeks. I was very pleased but, sticking by my promise to Steven and Lindsay, I flew back to New Haven without signing anything. I wanted to use the time between my trip and the signing of the contact to wait for any possible offer from Lindsay, but he offered nothing.

I called my board members to brief them on the results of my meeting at Water Pik and ask them whether I should sign the deal. Steven was very angry with me for not obeying his father and refused to discuss about my trip. The other board members (the lawyers) said that it was up to me to decide and that they would trust my judgment fully. Lawyers are the worst people to ask for advice on what to do. They will only tell you what "not to do." In other words, their typical advice is never constructive. I scratched my head and thought about the situation. On the one hand, here I had a sure thing—a deal with a reputable company, the thing I had been striving after for so many years. On the other hand, I saw a remote possibility of negotiating something with Jonson on his terms and working for him. But that, by any account, would not be a pleasant experience. Besides, Lindsay had offered me nothing but threats. So in fact, I had few real alternatives, and in the end I decided to go with Water Pik. When the contract arrived from Fort Collins, I again called the board members and told them that I was about to sign the contact. No one, not even Steven, objected, and I signed the document.

A week after the contract was signed, the first check arrived from Water Pik and I was able to start paying salaries to Ernie, Mac, and myself. Full of enthusiasm, we got down to work. The Water Pik engineers were to take care of the mechanical design, while our job was to make the special infrared sensor, design the electronic circuit and put together an advanced prototype. We were in a constant communication with each other, but we each still worked quite independently.

Within a few months, we began to experience substantial difficulties. The mechanical design Water Pik came up with was a straightforward approach, with little originality or any real concern for the end user. As conceived, the instant thermometer required a small metal plate covering the sensor which was to slide open for taking a temperature reading. The plate resembled the shutter in a photographic camera. The Water Pik engineers decided to use an electric motor to slide the shutter back and forth. This would certainly work, but unlike the simple design of a couple of springs and a plastic lever that I had used in my earlier prototypes, the cost of the motor and the supporting electronic circuit was far too high. Inevitably, this would raise the price tag of the finished product to a level clearly unacceptable for the consumer market. Nevertheless, the design had already been done, and we received from Water Pik all the mechanical components, put together the prototype with all its electronic guts, calibrated the thermometer and sent it to Colorado for testing.

A couple of weeks later I went to Fort Collins and was presented with the results of the testing, which had just been completed. They were not good at all. The sensor's output drifted too much and the thermometer did not function reliably when held by hand. I told my Colorado counterparts that I would have to take the prototype back to New Haven with me and analyze the problem.

During my visit, I was also shown the results of another part of the project: the design of the tools for producing the disposable probe covers for the instant thermometer. What caught my attention was that the new probe cover had many similarities to the covers produced by our rival, the IMS company in California, who owned a patent on their design. I asked the engineers why they had devised a cover that would likely infringe on somebody else's patent. In such cases, there are only two possible solutions. One is the design-around approach—that is, finding an alternative solution which is outside of the patented technology. Naturally, this is often very difficult and challenging. The other approach is to negotiate some kind of deal with the owner of the patent. For an engineer, however, the first approach is the only real option. Only if all alternatives have been exhausted can there be room for negotiation.

When I commented that the design might be an infringement, the answer I got was totally disarming. "We just do the design and let the lawyers worry about such things as patents. It's really not our business..." Apparently, it was just a job to these engineers. They really didn't care if the thing was successful or not. Once again I realized that a big company is the wrong place for developing an innovative product. There are no incentives and the people there are of a wrong breed for that kind of project.

Instant Thermometer

When I got back to New Haven and analyzed the prototype and the test data, it was clear that we had not one but several problems. The first was the sensitivity of the heavy motor assembly to the direction of gravitational force. Another problem was with the sensor I had designed. Unfortunately, my laboratory was very poorly equipped for producing a precision sensor with thin pyroelectric film. I had constructed it by hand, but it was a rather clumsy prototype and suffered from a number of drawbacks. Basically, my hand-made sensor was more complex than it really needed to be. The lack of appropriate equipment and the complexity of the design made everything far too difficult. After considering all the possible alternatives, I finally concluded that my engineering difficulties were the result of my own personality. My problem was that I have always striven to be as original and innovative as I could. Sometimes I have tried to be innovative for the sake of innovation, rather than for other, more earthly reasons. It became clear to me that both the mechanical design and the sensor design had to be changed, and that this change had to be one of simplification.

> Anything that is too clever, whether the design or the implementation, is bound to fail.
> Make everything simple.

I got down to work. Every day I was sending faxes to Colorado, and now and then I would fly out there for technical meetings. I had some ideas for improving the design, but I still suspected that this might not be good enough. What was even worse, I felt that the people in Colorado were beginning to lose faith in the product and were doing nothing to resolve their part of the problems, but were just waiting for me to give them all the answers.

We went on working. The sensor's design needed to be improved and I even knew what had to be done. But at the same time there was not the appropriate equipment either in New Haven or in Fort Collins to accomplish this, and what was even more important, I had no support whatsoever from my

Colorado counterparts. One day, Water Pik told me that it was going to stop making payments to FreMed until I could solve the accuracy problem.

I decided to abandon my original design for the sensor and use an analogous version which was already commercially available. The commercial sensor was of the same kind (pyroelectric) as I had been working on, and there was no departure from the spirit or the letter of my patent. The commercially available sensor was less elegant and slightly more expensive than the one I had initially wanted to design, but it had a few tremendously important advantages. It was already being produced in large quantities for various other applications, so there was no need to worry about manufacturing it and going through a learning curve. Also, it was readily available from several reputable vendors. While these sensors were being produced for other applications (mainly for motion detectors in security systems), I believed that I could easily adapt them for measuring temperature—which was what my patent was all about. The infrared ear thermometer was quite innovative in and of itself, and it made no sense for me to try to be too cute and include in its design additional innovations which were not essential to its function.

As soon as I realized that use of a commercial sensor might be an alternative solution, I looked in one of the trade magazines and found an advertisement for an off-the-shelf sensor distributed by Pace Electronics Inc. in upstate New York. I called the company, and its president, Patrick Kehoe, was at my doorsteps within a week. He was kind enough to bring a few free samples with him, and before another week was out Ernie, Mac, and I had tested them on the bench and discovered that the new sensor performed wonderfully. The lesson I learned from this was:

> Don't stumble on a difficulty.
> If there is no apparent solution, move along and get around the problem.
> Find an alternative approach.

By this time, we had already been working without pay for over a month, and I was delighted to be able to call Dave Smith and tell him that a solution had been found and we could continue as planned. Dave's response, however, was strange and very ambiguous, and I felt that he was hiding something from me. I called some other people in Fort Collins but still could get no meaningful response, except that we were not going to be receiving any more checks from them. It was a Catch-22 situation: without Water Pik's support we could not build the further prototypes needed to prove that we had a viable solution, but Water Pik refused to give in and pay us anything until we presented such a prototype. It turned out that this was just a tactic to get rid us.

I later learned that Water Pik now had a new president who felt that the instant thermometer did not fit into their strategic plan. He realized (quite rightly, as I now believe) that it had been a mistake to enter into an agreement with me in the first place. He decided that the company had to get rid of me neatly and cleanly, with no legal complications. But how to do this without breaking the contract? The solution they found was simple but insidious—slowly starve us to death. I found myself between a rock and a hard place. I was bound by the contract and could not go anywhere else with my patent, and at the same time I could not continue working on it for Water Pik. Indeed, it was a very innovative way for a large corporation to deal with an outside inventor.

Again we faced difficult times. I had to lay us all off, and as before, the state of Connecticut refused to pay me unemployment compensation. Ernie decided to stay and went on working without pay, but Mac left and took a job with the post office. It was the summer of 1987, three years since I had started FreMed, and I began to realize that the enterprise was not working out. I felt that I had done everything I possibly could, but I also had to think about my family. By this time I was living on credit cards and my debt was growing alarmingly high. I even called myself a termite—I had to take a second mortgage and we, so to speak, ate our own house. Irina supported the family as best she could, but on a musician's pay this was feasible in theory more than reality. When I realized that we had no money for our daily expenses, I began looking for a job.

During those difficult months I still went in to the laboratory (which Science Park continued to allow me to keep without paying rent—what would I have done without such good people!) and sometimes I worked at home.

Since there was little more I could do on my instant thermometer for now, I worked on some other ideas I had. One of these was another innovative approach for measuring temperature without touching an object. I believed that I had invented what was at that time a new generation of infrared thermometry, going beyond even my previous invention. I composed a patent application myself, since I could not afford a patent lawyer, and filed it with the U.S. Patent Office. Soon, I received a letter from the examiner telling me that this was the most original invention they had seen in years (it is quite unusual to receive such a compliment from the Patent Office) and the patent was granted shortly afterward with only very minor corrections. I decided to call Water Pik to tell them that if they did not like my old approach, I now had a better idea, but they flatly refused even to discuss it.

When I realized that there was no light at the end of the tunnel (if you see no light at the end of the tunnel, you are probably looking in the wrong direction), I picked up the telephone and dialed Steven's number in Cleveland. "I'm sorry Steve," I said, "but it looks like our enterprise didn't work out. I think it's time for me to close the shop down for good and take a regular job."

Ten minutes later his father called.

Lindsay Jonson said that he might finally consider taking the whole thing over himself, but only on his own terms. Before talking terms, however, he wanted to evaluate the patent and the market potential one more time. He asked his friend John Trenary to find independent reviewers to give him their professional opinion of the invention. At the time Trenary was president of Atlanta-based Dental Research Corporation, which produced the Interplak electric toothbrushes. When I had started FreMed, Jonson had acquired the patent rights to that invention from a Chicago inventor, George Clemens. He had hired Trenary from Water Pik (it's a small world, after all) to run the newly formed company, which in a few years they built into a multi-million-dollar business. Shortly after my relationship with Water Pik reached the stagnation point, Jonson and Trenary sold Interplak to Bausch & Lomb for a hefty 133 million dollars. Both of them had received heaps of cash from the deal and were obviously looking for a good place to park this money, preferably at a large profit. From that prospective, my instant thermometer was definitely an option.

Trenary had a degree in mechanical engineering, though his engineering skills were all but non-existent. Still, he always pretended to have the ultimate knowledge in everything, from electronics to marketing to finance to sales. He also had a degree in business administration, and managed the toothbrush business according to the textbooks he had recently studied: technically correct, but with very little imagination or creativity. I suppose he had an unconscious and unfulfilled dream of becoming a general, because he loved nothing more in life than issuing orders and hearing other people reply, "Yes, sir!" That was the style he imposed on Dental Research Corporation, where everyone was only longing for the time that obnoxious man went someplace else. Eventually, this would happen, when the company was sold, but at the time Jonson asked him to evaluate my invention, Trenary was still in Atlanta.

Lindsay Jonson, Steven and Trenary arrived in New Haven together, met somewhere in town and then had only a brief talk with me. They used every possible means to show me that they were now the ones in total control and that my position was only one for saying, "Yes, sir!" Yet they were deluding themselves, for no matter how they viewed my position, I was still the same person as before. My persistent independence irritated Lindsay Jonson a great deal, and he decided to communicate with me primarily through his son Steven and Trenary.

After a few weeks, they hired two patent law firms to evaluate the strength of the patent. When both firms gave them a positive response, Trenary told Jonson that it made sense for them to invest. FreMed already had too many shareholders, whom Steven had drummed up to raise money for us to stay afloat. However, Lindsay Jonson wanted to have the unconditional control over the entire business, so he developed the idea of starting a new company and buying out all rights to the invention from FreMed and Water Pik. At this point the real negotiating began. In effect, it was between Lindsay and me, yet it was Steven and I alone who discussed all the details over the telephone. Clearly, this was a classical case of a conflict of interest, since Steven, though a director of FreMed, was negotiating a deal with FreMed on behalf of his father. Lindsay could not forgive me for my decision to go to Colorado and make a deal with Water Pik against his express command. His hatred toward me was matched only by his son's. However, Steven's business talents were a far cry from those of his father and the only unsurpassed skill he possessed was being a fast talker.

The initial offer they made me was nothing more than a joke. They proposed buying out the contract with Water Pik, and paying FreMed (of which by this time I only owned about 25% of the shares) a royalty of one half of one percent, and that only after deducting all their expenses. This would have meant that I was virtually donating both the patent and the know-how to them in exchange for a tiny amount of money that I might possible receive sometime in the remote future. I turned them down flat.

Steven was furious. He screamed so loudly that I had to hold the receiver away from my ear. I made him my counter-offer, one which I felt to be a fair deal, but my opponents just laughed. This pinball-like telephone negotiating went on for several months. In financial terms, I was broke, and Ernie had no income either. Fortunately, in order to stay afloat I was able to secure a small loan from the City of New Haven, but that lasted for only a short time.

What I could not comprehend was why Jonson was so greedy with respect to the royalty percentage. Between himself, his son, some relatives, and his friends, they owned a majority of the FreMed stock and the biggest portion of these royalties would be going to them anyway. So why then were they being so stubborn on that particular point? Most likely, the desire to humiliate me and take whatever they could from me was far bigger than their natural greed for money. Even Jonson's own lawyer Mark Kaufman muttered to me at one point, "Well, sometimes he gets carried away."

In November of 1987, the three of them and I all flew to Fort Collins to negotiate the buy-out of the patent rights with Water Pik. I tried to convince Jonson that Water Pik would be quite happy just to get rid of this project and that we could get the release papers almost for free. But again, just for the sake of not agreeing with me in anything, Lindsay told me to keep my mouth shut. Though he could have gotten the rights for nothing, he ended up paying to Water Pik a quarter of a million dollars. But this did not bother him, for this was money he expected to get back by subtracting it from the royalties. After that visit, it took another six months to draft the documents, and in March of 1988 Jonson incorporated a new company which he named *Thermoscan*. That company acquired all rights to the invention from FreMed in exchange for paying off its debts and hired me as vice president of research. Trenary was appointed as acting president.

According to my employment agreement, I was allowed to buy a small number of stares in the newly formed corporation. Sixty percent of these shares, however, would be vested only on condition that I produce a working prototype of the instant thermometer. The prototype had to be built according to specifications included in the contract. If I failed to do this, I would lose all rights to those stocks. I thought that this was fair, for if I could not get the thing to work, who would need those stocks anyway?

On April 12, my birthday, I was officially hired by Thermoscan, though I continued to work at the same laboratory in New Haven, which received a fresh sign on its door: "Thermoscan Inc." I decided to build an infrared thermometer without any regard to the cost of the prototype. At this stage that was not all that important, though. We knew that if it worked, the commercial version could be produced relatively inexpensively. What we had to accomplish within the next six months was to demonstrate that our technical solutions were sound and that the thermometer could measure temperature instantly with an accuracy of better than one tenth of a degree.

It was an intensive, but very productive period. For the first time in five years I was able to do what I did best. I was no longer having to raise money; all I had to do was use my engineering skills. Before the end of August, we had built two prototypes. All the mechanical and optical parts were fabricated by the precision machine shop of Yale University, a superbly equipped place with highly skilled machinists. The cost of the parts to us was about a thousand dollars per prototype—not terribly much given the excellent quality and the experimental nature of the work.

We gathered about fifty people from the various start-up companies at Science Park and measured their ear temperature as compared to a reading made with a conventional oral thermometer. To better collect and analyze the data, we connected our prototype to a personal computer. The test results were very good. The prototype functioned reliably, with an accuracy to within a few hundredths of a degree—far better than we had expected.

I prepared a report and sent it to Trenary. In the cover letter, I mentioned that I had fulfilled my promise and even exceeded the requirements. However, in the course of our research, the specifications had had to be slightly modified. In particular, instead of the nine-volt battery planned for the commercial unit, we used a line-operated power supply, as that simplified the interface between the prototype and the computer we had used to collect data. Of course, it would be very easy to convert it back to a battery version in designing the commercial product.

Trenary responded that he was satisfied with the results, yet because I had violated the specifications by not using a battery for the prototype, I would lose my right to buy any Thermoscan shares. That kind of meanness I had not expected even from Trenary. I called Lindsay and Steven—but they all said, no, you've lost your stock. Clearly, this was just a part of their revenge for my not getting off that plane two years earlier. I was very angry and tempted to slam the door right in their faces, leaving the lot of them with a couple of prototypes which without me would be good for nothing other than cracking nuts.

I called Alan Bartholdy to ask his advice. He said, "Don't do anything irrational. Nothing is carved in stone. Maybe one day you'll be able to persuade them to give the stocks back to you." I thought about it and decided to bide my time, for I already had a trump ace up my sleeve.

You can get more with a kind word and a gun
than you can get with a kind word alone.
-*Al Capone*

Lindsay Jonson was a most unusual man, possessed of a number of qualities to which one might attach the adverb "very". He was very smart, very rich, very snobbish, and very obnoxious. After he made all that money with Interplak, he became so proud and full of himself that he thought whatever he touched would turn to gold and every word he uttered must be heard with admiration and perceived as holy revelation. In spite of such a repelling attitude, he had some admirable and quite unusual distinctions rarely found in people of wealth. Unlike most of his counterparts with large bank accounts, he did not play golf, did not gamble in Vegas or engage in idle small talk at endless social gatherings. Most of all he liked adventures with risk and danger. He traveled all around the world, crossing New Zealand and most of Europe on his mountain bike and climbing Kilimanjaro in Africa and Elbrus in Russia. Also, he was inclined to do something that most other investors would never touch with a ten-foot pole. He liked to invest in high-tech inventions and build new businesses from the ground up. Eventually, he was successful in both: he enjoyed his travel adventures and made a lot of money with his investment enterprises.

When he acquired the rights to Interplak and announced that he was going to market a new electric toothbrush to consumers for $100 apiece, everyone laughed. Who needed another toothbrush? How many of them had already been designed since the first electric version was patented in 1885 by the American inventor Scott? Yet Jonson said that the new electric toothbrush would not be just a toothbrush, but a "plaque-removing tool." He started a company, hired people who knew the toothbrush business (which is how Trenary got the job), and secured professional validation through multiple clinical studies. These studies proved that Interplak was indeed good for removing plaque from the tooth's surface. And also important, he was able to convince millions of consumers through clever marketing and PR campaign that the new toothbrush was worth a hundred dollars. He planned carefully, calculated the risk, and assembled a skillful entrepreneurial team for the commercialization of the Clemence's invention—all of which resulted in a great success.

That is why I have mixed feelings about the man. I both hate and admire him at the same time. I hate his arrogance and snobbishness, but I admire his bravery and entrepreneurship. I had searched for a backer for my inventions for many years, and I had even found several who initially agreed to go along with me, but few of them had the guts to finish the journey. Jonson was the one who had the strength and determination to accomplish this difficult goal. Yes, he made a lot of money with it, but I believe he deserved it. I just wished that he were more human.

Once I proved to him that the instant thermometer was indeed instant as well as an accurate and inexpensive instrument, he followed his own footsteps. He did the same thing he had done with the toothbrush: engaged an executive head-hunting firm to search the entire country for top-notch people with expertise in medical thermometry and consumer marketing.

The first person he hired was Bob Lackey, who worked as a director of R&D at IVAC, a San Diego medical instrumentation company. Bob's wife Phyllis said that under no circumstances would she leave California, so Jonson said, okay, let's set up Thermoscan in San Diego. Bob began working out of his garage on the West Coast, while I kept on working in my old lab on the East Coast. The company made its first steps toward the production of a commercially acceptable product.

When Thermoscan took off, New Haven Science Park proudly announced that I was one of its success stories. The people there supported entrepreneurs like myself and sincerely rejoiced in my success. They even asked a local newspaper to publish a piece about me. The article began with words, "His name's Jacob Fraden; Robin Williams will play him in the movie..." Apparently, the writer, Paul Bass, intended to express in allegorical terms that I reminded him of a character from the movie *Moscow on Hudson*. My children, however, took his words literally and, I think, for the first time looked at me with respect.

Through all this, the friction with Lindsay Jonson continued. It seemed that he had developed a habit of listening to what I said but then doing just the opposite. For example, I suggested that we design the instant thermometer for the professional market first, that is, for doctors' offices and hospitals. Only then, and only after we had passed through the learning curve, should we design and produce a less expensive consumer model. I also proposed that we hire a group of mechanical and electrical engineers and do the design in-house. Instead, he subcontracted two companies to develop both models in parallel, the professional and the consumer. After two years and the loss of several million dollars, everyone came to realize that this had been a mistake. The subcontractors cost the company many times over what three or four full-time engineers would have done. Moreover, such engineers, being our own employees, could have accomplished the goal much faster. But since we had to work on two thermometers and a new disposable probe cover design all at the same time, everyone ended up being spread so thin that the consumer model was not finished on time anyway, and we had to introduce it to the marketplace a year after the professional version. But of course, Jonson never admitted that he was wrong.

> **Use subcontractors and consultants for technical work only. Don't count on their creativity.**

Since Trenary was still in Atlanta, Lindsay hired a new president to run the company, Bill Krookel, a man who possessed two qualities. First, he was far from bright, and second, he was a brown-noser and a yes-man, something that Jonson liked in his subordinates. As a company president, the man was a complete joke. He could not even decide what color carpet to put down in his office without first consulting with Jonson.

During the first two years of Thermoscan's existence, I commuted nearly every month from New Haven to San Diego, spending one or two weeks there each trip. The further along the engineering work progressed, the more often my presence was required in California. Upon arriving at the Ramada hotel were I usually stayed, I greeted all the maids with, "Hi, honeys, I'm home!"—to which they replied, "¡Hola, señor!" Soon,

Instant Thermometer 329

these trips had become quite inconvenient and expensive, and the engineering people pressed Jonson and Krookel for my permanent relocation to California, saying it was essential for our further work. I agreed that it was in the best interest of the business, but it was not an easy decision to move from Connecticut to the West Coast. I heard before that California was a bit unusual place where people had a different mentality and easy-going attitude. In other words, it was full of "fruits and nuts". Somebody told me that for one year spent in Southern California one must deduct one point from his IQ.

So I decided to attack this problem scientifically. I called MENSA and asked them to send me a standard do-it-yourself IQ test. When I got the package, I did the test and sent it back to MENSA. In a few weeks I received the letter with the results. My IQ according to the test was 143. At that time I was 45, so I figured it should be about twenty years till my retirement, meaning that at that time I still will be safely away from a mentally retarded state. After that relieving news Irina and I decided to make the move. We sold some furniture and put our Connecticut house on the market, Irina resigned from the New Haven Symphony Orchestra where she had played for eleven years, and in summer of 1990 I moved to San Diego. Irena and the children joined me a few months later.

Soon after my relocation, Jonson sold more shares of Thermoscan to a group of investors. All the key employees were allowed to buy some stock as well. He felt, quite correctly, that shares in a venture were a strong incentive and a reliable driving force for making people work harder and accomplish what was expected of them. Everyone who wanted them was getting shares—except me. He did not allow me to buy any.

> **Share tomorrow success with your employees and they become your best partners today.**

I realized that the time had come for me to take action. I called Jonson and demanded that he allow me to buy at least the shares of Thermoscan that had been promised to me at its formation and then viciously taken away. Jonson flatly refused, saying that the deal with me had been closed two years ago and that he did not want to discuss it. I asked why every V.P. of the company was allowed to buy a significant number of shares, everyone except me. His response was more than blunt.

"Because they are high-level professionals and to get them here and make them work hard I have to give them these lucrative incentives. But you, I already have you here. You are an inventor, you would stay and work no matter what, just to see your brainchild grow. Why, then, do I have to give you anything?"

I was disarmed by his honesty, but still curious.

"Why are you doing this, Lindsay? Surely it would be in your interests to keep me happier here. Besides, the number of shares that I was entitled to buy is a microscopic fraction of one percent of the company. It is so minuscule that nether you nor anyone else would even see the difference. But for me it's a big deal. Why are you so stubborn?"

"That's the way I am," was all he said. Some people to enjoy a good reputation, give publicly, and steal privately.

I had no choice but to pull out the trump ace from my sleeve.

"Lindsay," I said, "I want to let you know that a few years ago I invented another thermometer, better than the one we make now. I have the patent already issued, and if you won't allow me to buy the shares to which I am entitled, I'm quitting my job and going somewhere else with my new patent."

It was like a bomb blast. Jonson was furious.

"All your inventions belong to us! To me!! You've been working for us all these years! Read your employment contact!"

"Not quite so," I said. "I invented this new method well before Thermoscan was formed and I was living at that time without any pay from either FreMed or Water Pik. I was on my own, all my steps were properly documented, hence, I am the sole owner of that patent."

Jonson then proceeded to threaten me in the manner so common of people of his caliber. "I'll sue you!"

"Do what you want." I had made up my mind to fight him to the end. "But you know darn well that no court in this country will find me guilty of anything."

"I don't have to win in a court! I can afford the best lawyers in this country. The best and most expensive! Can you do the same thing? You can't afford to fight even a few days with me! You'll be broke long before the case goes to trial, you'll lose your house, your car—everything you have. Think of your wife and children! I'll throw you and them out, out on the street! Take my advice, give up now and transfer all rights to that patent to us!"

As I said it before in this book, nowadays a "civilized" mugger attacks you with a lawyer, instead of a knife. I refused to talk to him in such a manner, and so he let his pack loose on me: Ralph Lilore, Bill Krookel, his son Steven, and his lawyer Mark Kaufman. At first, they spoke sweetly and amiably to me, then viciously and crudely. The worst of the gang was Steven, but I stood firm.

To be fair, I must mention that Lindsay and his son were not unique in their greed. Before them and many times after, I met a lot of people whose business was the money making. A great majority of them had passion for ripping off inventors as viciously as they could and at any possible occasion. Obviously, such people do not have much appreciation for creative minds and the vital contribution an inventor brings to the innovative business. They do not realize that a good idea does not pop up by itself—usually it grows in inventor's head from many years of studying, work experience, and hard labor. Life of an inventor is far from being easy, and it is pity that men of money don't regard him as an equal partner, but rather as an object of their standard *grab-as-much-as-you-can* tactic.

One day Lindsay told me, "Everything I do in my life must meet two requirements. The first—I have to have fun with whatever I do. The second—I have to make money with whatever I do."

"Then you are doing it wrong, " I replied. "I'm sure you'll have a lot of fun screwing me over, but there is no way you can make any money with it."

Another time I said, "Be reasonable, all this commotion makes no sense for you. Let's assume for a moment that you prevail and take this patent away from me by force. Then what? Without me it is just a worthless piece of paper. The only person in the whole world who knows how to make this new invention work is still me. Me alone. This country has abolished slavery and there's no way you can force me to work for you against my will. You can steal my patent, but you can't steal my brain. Your victory will turn into your defeat. You'll lose both, my new invention and this company too, which without my participation will very likely go haywire. Think about it and let's start talking, not fighting."

That kind of argument brought him to senses, and little by little his position softened and we began negotiating. Things would have gone far more smoothly, though, were it not for his son Steven, who inherited the worst aspects of his father's personality. All too often, after long and exhaustive negotiation, Lindsay and I would reach an acceptable formula, but when I received a draft of the agreement from his lawyer a few weeks later, everything would be twisted around. It turned out that Steven did not want to make any compromises with me. He wanted to get everything for nothing, so he and their lawyer kept changing the agreement by sneaking time bombs and land mines in it here and there, in the hope that I might neglect to read each new revision through and carelessly sign it. My financial position was not all that good and I could not afford to have my lawyer review each draft.

In spite of the high pressure from my opponents, I stood firm and demanded that in exchange for the rights to my new patent, I be allowed to buy the Thermoscan stock to which I had been initially entitled. Besides, I wanted to negotiate different royalty terms for the new patent, terms more favorable than those I had been forced to accept two years earlier for my original invention. This demand infuriated Lindsay and Co. even more.

While negotiating with Jonson, I remained tenacious and resilient, and finally, after long nine months of exhausting talks, we signed a deal which in most respects I considered reasonable. I was allowed to buy a portion of shares that have been promised to me two years before (alas, I now had to pay a much higher price for them) in exchange for licensing my new patent to Thermoscan on somewhat better terms. The price I paid for those tough negotiations, was the deterioration of my health. My blood pressure skyrocketed and I had to go to four different doctors before I found one who was able to get it under control. Sometimes I ask myself, is it worth it?

December 4, 1990, was a very remarkable day for me, because that was the day Thermoscan shipped the first instant thermometer. Actually, it was a big day for several other people who had worked day and night to make it happen. Besides the workaholic Bob Lackey, engineers Bob Bohl and Joe Brown all made a number of crucial contributions to the design of the thermometer and probe covers. Betsy Winsett, our vice-president of marketing, created a clever and innovative marketing campaign, which was backed up by the extensive clinical research organized by the vice-president of professional relationships John Hyle, a man with a sharp mind and a great sense of humor. There was no question that Lindsay Jonson put together quite a team.

Betsy developed a concept of "Thermoscan Kids"— a number of funny looking children with sad faces (there are plenty of happy kids in all sorts of commercials and nobody pays attention to them anymore.) The first sad baby was named Timmy Panic (after word *tympanic*), another was Annie Eartug (to stress a better measurement technique—tug the ear when taking temperature), and then was Barry Accurate (*very accurate*.) The children were sad because they objected the old-fashioned way of temperature measurement, namely—rectal. These kids were used in all Thermoscan advertisements and soon became well recognizable by the general public, like Gerber baby or Pinocchio.

We sent out the first shipment of thermometers, everything was in place, yet the company started falling apart and we found ourselves on the verge of bankruptcy. Thermoscan looked like a collection of beads in which all the pieces were precious and beautiful, but useless for making a necklace, for there was a link missing. The problem was that our president Bill Krookel, whose talents were insufficient for the running of a hot-dog stand, let alone a high-tech medical instrumentation company, could not make a single decision without first getting permission from Lindsay. In effect, Jonson was the one who ran the company, whether from his home in Cleveland or his retreat in Vail, Colorado, and he did it through Krookel quite poorly.

> Allowing an accountant or lawyer to run an innovative business is a sure way to failure

When the bank account had dried out, Lindsay realized that he had to get rid of Krookel. It was clear that he should have done it at least two years earlier, but his inability to admit his own mistakes brought us to the edge of the abyss and cost many millions of dollars. Apparently, Lindsay did not know too many talented businessmen crazy enough to work for him. And so, as he had once before, he offered the job of president to his old buddy John Trenary, who by this time had left Interplak with the hope of retiring to the golf course. But Jonson offered him extremely lucrative incentives to take the company over, and Trenary accepted.

Even though I despised Trenary, we were all so fed up with Krookel that any other choice looked like a blessing. And indeed, Trenary managed to pull the company out of the hole and establish some financial discipline and communication between the various departments. Alas, the company suddenly shifted from a total anarchy to a military-style enterprise. As at Interplak, Trenary ran the business according to textbooks and liked nothing more than to hear people say to him, "Yes, sir!" A text book approach may be okay, as long as it is a good textbook and one doesn't get too carried away with it.

Trenary made a lot of changes in the company, some good, but many more just for the sake of change. His senseless administrating caused a great deal of irritation and created an unhealthy work atmosphere. He fired the company's lawyers, accountants, travel agency, insurance providers, consultants—everyone. The reason? He just wanted to show that he and he alone was in ultimate and total control. George Bernard Shaw once wrote about a conversation between Napoleon and his aid:

Napoleon: What shall we do with this soldier, Guiseppe? Everything he says is wrong.

Guiseppe: Make him a general, Excellency, and then everything he says will be right.

Still, strong financial discipline and adherence to a business plan, a strict policy of checks and balances, were right and productive moves. Investors liked the difference Trenary made to the balance sheet, but the company employees hated him and could not wait for the glorious times he flew back to Hawaii, where he had built a house and wanted to settle for good. One day all employees of Thermoscan heard some good and some bad news. The good news was that Trenary's brand new home in Kauai had been demolished by a hurricane. The bad news was that he had survived and that now he planned to spend more time in San Diego.

One year after we shipped the professional model, a consumer thermometer was also introduced and initially sold through the yuppie catalogues and stores of *The Sharper Image*. It soon became their hottest selling item. Later on, the instant thermometer appeared in every imaginable class of trade: from airline mail order catalogues to mom-and-pop drugstores. Our V.P. of sales, Pete Ellman, was a magician of trade. He was the first man to teach me that selling could be elevated to the noble level of an art.

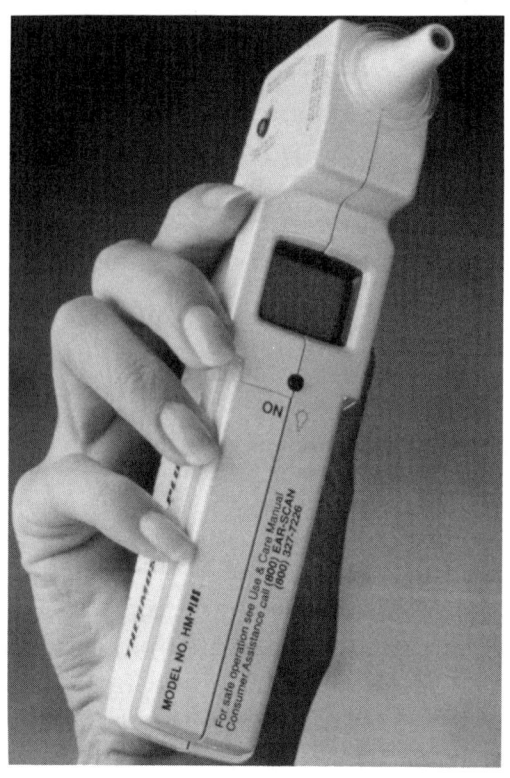

Consumer model of Thermoscan Instant Thermometer

We had three plants—one in San Diego, one in China and one in the Philippines—producing thermometers round the clock. Nearly everyday large crates of the consumer-style instant thermometers arrived by sea or air to be distributed from our warehouse in San Diego. Another plant in Colorado produced disposable probe covers. In March of 1994, on Thermoscan's sixth anniversary, we shipped the millionth instant thermometer. Indeed, our slogan, "We change the way the world takes temperature," had a lot of truth in it[1].

For the first two years of sales, all royalties for the invention went entirely to Lindsay to cover his cost of buying out the FreMed assets. But finally, more than eight years after the

---

[1] In 1995 alone Thermoscan shipped nearly 2 millions instant thermometers

initial conception of the instant thermometer, I received my first check. It was small, but naturally it meant a lot to me. Later on, the sales grew and so did the royalties.

In spite of the agreement allowing Thermoscan to use my new patent, Lindsay & John did everything possible not to use it. It was obvious that the new technology would be very beneficial and the company could profit from it tremendously. Nevertheless, their desire to punish me for my "disobedience" was so overwhelming that they mothballed the invention. For nearly five years those two avengers preferred to shoot themselves in the foot, refusing to fund the project or allocate any of the company's employees to work on its research and development. I knew that such a practice was illegal with respect to a licensed patent and at one point even considered taking some sort of court action, but later I decided not to jeopardize the company's business, which had started to gain momentum.

My job at Thermoscan became quite tedious, for without support from the company, my work on the new project came to a standstill and there was nothing I could do to develop the invention further. Nor could I work on any other ideas not directly related to the company business. I could not even quit, as that would violate my employment contract and I might lose everything I had achieved during all those long years.

> Before you sign any contract make sure that it contains clean and safe exit for yourself

I performed a lot of support work for other departments and worked on a computer with some mathematical models for the future thermometer, hoping that one day I would be able to construct it in hardware rather than software. Yet I longed to do something more tangible than computer simulations.

To keep myself occupied in a creative way, I again turned to painting. I have always loved to paint. It is so much like inventing, only easier. When you invent something, you have no choice but to make it work—not only on paper, but in real life as well. Inventing is a great challenge, because there is only one way to do it right—the way that doesn't clash with the laws of Nature. In painting, meanwhile, you do what you want how you want. With brushes and paints, I can allow my imagination to run wild, against all the laws of man or God, and it's all perfectly okay as long as the work does not conflict with good taste. That is why I like the surrealistic style—it allows me to invent without limitation any image that comes to mind.

Unfortunately, I never studied painting professionally and have often felt that I lacked a fine technique. I believe that mastering a trade is equally important in science, engineering, and art. I do not like most modern abstract paintings. I cannot shake the feeling that abstract painters are hiding their lack of skill, talent, and imagination in a pretentious abracadabra of splashes of paints in the hope that no one will have the courage to say, "The Emperor isn't wearing any clothes!"

The great surrealists, however, offer to us a marvelous synthesis of imagination, feeling, and fine skill. When I see how the imagination of Salvador Dali raged on canvas, I envy his brilliant and masterful painting technique, even if sometimes I do not appreciate his paranoid Freudian vision. I feel much closer, though, to René Magritte, the Belgian surrealist. Living in Russia, I never heard his name (the surrealism was a forbidden art), and when I began painting in America, somebody who saw my pictures said to me, "Oh, I see you are copying from Magritte!" Curious, I went to a bookstore, found an album of Magritte's paintings and discovered that, indeed, without knowing it I had repainted some of his works almost exactly. But of course, Magritte did it much better.

Besides inventing and painting, I like teaching, yet that became next to impossible while I was working in industry. So I decided to write a book about sensors in which I could teach a reader who was interested in that subject. I had found that writing a technical book was like talking to an imaginary pupil. The book I wrote was initially conceived as a textbook for graduate engineering schools, but later I changed it to a handbook. It took me eighteen months to put the book together, and in 1993 it was published by the American Institute of Physics. It was titled *Handbook of Modern Sensors*, and after a few months in print it became a bestseller and one of the most profitable books by that publisher.

At about the same time I was working on the handbook, I was invited to become the chairman of a committee developing a new National Standard for medical infrared thermometers. At first, I thought that writing a standard would be quite a boring job, but later I became absorbed by the almost mathematical grace of that work. Writing a standard is like writing a law— it must be constraining and at the same time permissive, and it must be precise as well. The life span of a technical standard is relatively short, five to ten years at most, for as technology progresses, a new standard must be developed. While working on the standard and trying to strike a balance between "may" and "may not," I constantly admired those great minds who developed and wrote the Constitution. That document was a kind of political standard for the nation, a standard that has not only survived well over two hundred years, but is still the best constitution any country ever had. Quite a standard indeed!

Beginning on the very first day of 1994, Thermoscan offered itself for sale. Lindsay Jonson and the other large investors wanted to cash in on its success. It's the game they play— bring the company value up, sell it, and go to something else again. We, the entrepreneurs who had worked hard for all those years, also wanted to make some money from the deal and propel the company onto a higher level. We knew that whoever invested money in Thermoscan would make sure that our present achievements became translated into long-term success.

The company engaged an investment banker and prepared an offering memorandum. Soon, seven suitors lined up. We all, and especially the marketing and financial people, were required to make numerous presentations to convince the potential buyers of the high value of Thermoscan. Over the past several years the company had grown so successfully that it was nearly doubling its revenues every year, surpassing sixty-million in sales in 1994[1]. Obviously, its price tag rose quite high, and within a few months, five of the "grooms" dropped out of the race, unable to compete. The remaining two intensified their efforts to take over the company for a price they could afford to pay and the sellers would agree to accept. Finally, just one buyer was left, Thomas H. Lee Co.—the investment group from Boston. A leveraged buyout was planned, with Thermoscan to be subsequently put on a public market. Many of us liked the idea of trading our stocks on a major stock exchange, because we knew that it would be one hell of a public company! We already had lots of ideas for introducing new products and expanding business into the international markets.

Thermoscan was changing hands and everyone hoped that it was a move in the right direction. Trenary and Jonson were ousted and the entire company had changed dramatically—it became more dynamic and flexible in operations, more aggressive in the market place, and what is very critical for a long-term success, a much more pleasant and rewarding place to work. We looked forward to changing the way the entire planet takes temperature, because we knew that Thermoscan was the only company in the world capable of doing that. Apparently, such high marks to the Thermoscan performance have been given by some other people as well. Just two weeks before the initial public offering, The Gillette Corporation made a bid and bought out all Thermoscan stocks for nearly 190 million dollars. Gillette is one of the U.S. largest companies with over 6 billion dollars in sales and excellent reputation for quality and innovative thinking. That unex-

---

[1] In 1995 the annual growth rate was 140% and sales reached $113 million

pected turn of fate wide opened doors to the world markets and the Instant Thermometer was given a new and powerful momentum.

I am looking back at twelve years of joy, frustration, and struggle, and I proud of them. The job was long and difficult, but it was worth it. The invention became very successful, Thermoscan grew to become the largest infrared thermometer company in the world, giving jobs to thousands of people in America, Europe, and the Far East. Our key employees achieved financial security and investors received very hefty return. During those years I did many things right and made a lot of mistakes, and if I had it all to do again, I would surely do it differently. But as a Russian proverb says, "If I had known where I was going to fall, I would have put a rag there."

Appendix

Making Inventions

Character of an Inventor

> All you need in this life is ignorance and confidence, and then success is sure.
> - *Mark Twain*

A dictionary defines the word inventor as "one who contrives a previously unknown device, method, or process." The Patent Office would probably disagree with such an interpretation, and for good reason. Not everything that was previously unknown is an invention in the legal sense.

The greatest majority of patents granted by the U.S. Patent Office protect innovations conceived by employees of large corporations. There is no doubt that in most cases such patents are issued to genuine and generally useful inventions. But these inventions are accomplished in the line of duty, their authors rewarded, at best, by a small bonus and a handshake from the boss. Rarely is there any great emotional or monetary driving force behind these inventions. Many corporate patents protect small improvements and are intended to keep the competition at bay. Corporate inventors put these patents on their resumes and feel good about them, and rightfully so. But most often, their inventive function ends with the filing of a patent disclosure and the signing of some legal forms. The rest is corporate business.

By contrast, the life of an *entrepreneurial* inventor is quite different. For him, the filing of the patent disclosure is just the beginning. The inventor-entrepreneurs are individuals, or perhaps small groups of independent entrepreneurs, who go through the whole process of making it happen. They not only conceive their dreams, they also bring them to term, deliver them, nurse them, and raise them to adulthood. Inventing is very aptly likened to making a baby: it's fun to conceive but hell to deliver.

An invention is not the logical result of putting together some common facts or ideas. Rather, it is an illogical twist, a kind of mental singularity. In order to produce an invention, the thought process must go through a process of revolution rather than evolution. An inventor is someone who is capable of performing this type of uncommon work. In addition, my own definition of an inventor is narrower than that found in the dictionary. It includes specific characteristics of personality and human behavior, because an inventor, in my view, is someone who is on a perpetual mission or quest for a dream. For a few lucky ones that dream comes true, but for many others it remains an elusive goal forever.

The inventor who has the strength and determination to fight for his brainchild can sometimes make real money, which is something altogether different outcome from just getting a patent, framing it and hanging it on a wall. The financial reward is, indeed, a strong incentive for going for it. I must add, however, that inventing is not a way to get rich quick. If making money is your only goal, one should look for faster, easier, and more reliable ways of making a buck, such as gambling on a horse race or pulling the lever of a Vegas one-armed bandit.

The casualty rate among inventions is enormously high. No more than one out of every hundred patents earns enough money to pay back development costs and patent fees. And perhaps one in five hundred makes any money beyond its out-of-pocket costs. In a moment, I will talk about cashing in on your idea, but here I just want to stress that inventing, like any creative profession, is only for those who truly cannot contemplate doing anything else. If you feel at all comfortable in any other less demanding and less stressful profession—then do that. The inventing business is too dangerous for those who are not ready for a long struggle and who cannot accept the risk of failure.

Appendix: *Making Inventions*

> Professionals built the Titanic, amateurs—the ark
> - *Frank Pepper*
>
> If I had thought about it, I wouldn't have done the experiment. The literature was full of examples that said you can't do this.
> -*Spencer Silver* on the invention of the 3-M 'Post-It' notepads

An entrepreneurial inventor must possess very special qualities. To be sure, that is true for any profession. But for an inventor to be successful, it is essential to combine quite different and at times mutually contradictory values.

Making an invention is an art of creation. Far from being a science, it is rather a highly individual and intimate process, much like painting a picture or composing a poem. For many inventors, the creative process is a natural outcome of their personalities. Claude Monet, the great French impressionist, said, "We paint as a bird sings." If one asks how lyrics are written, a bad poet (if the words "bad" and "poet" can be placed next to one another) will explain the mechanics of writing. He may describe the rules of rhyme or define measures and styles. He may dissect a poem, like a frog on a laboratory bench, and teach you the trade of "poeting." But a *real* poet will just smile at the question, for it is far too intuitive and intimate to describe in logical terms. It is the same with inventing.

Yet poetry does go scarcely beyond the act of writing. When verses are written, that's about it. The job is basically done. It is either good or bad. Either people like it or they do not. To understand and feel poetry, you must possess just two essential qualities. You must know the language and you must have a heart that is open to the lyrics. For an inventor, however, putting his idea down on paper is not nearly the end. It is but the first step on a long journey.

To produce new stuff, inventors and poets need inspiration, just as cars need fuel. What pity that you can't buy inspiration at a gas station! It is something mysterious, an act of God much like an earthquake or a tornado. When it comes, it can turn everything upside-down. But unlike a tornado, inspiration does not appear by itself out of the blue, nor can you will it to come. The problem you are trying to solve must first be nurtured, it must develop and grow until it reaches a critical mass, when a solution suddenly flashes through. And that is the moment of inspiration.

If you want to invent something, set your goal and go at it a day, a month, a year, perhaps much longer, as long as it takes. Time may pass and it doesn't work. Something may not fit together. You become frustrated, while your wife asks why on earth you don't take a regular job like other normal people. But you keep going, because you believe that sooner or later you will find the answer. Like a rain man or a witch doctor, you strive for that flash of inspiration, you pray for it. You become pregnant with your idea and you must carry it to term and bear a healthy baby. Your soul and mind must be fine-tuned to it. Time and again you hate it and think, "To hell with it! I'll do something easier. Like normal people do." But then you return again and think about it some more. Day and night. You have dreams about it. You may go to a concert or read a novel or play some ball, but it is always on your mind. It gnaws at your brain like a disease.

And then, suddenly—bam! You have it. Often, your solution seems so obvious that you scratch your head and wonder why on earth you never thought of it sooner? Or for that matter, why nobody else before you thought of it either? You turn your idea over in your mind and see that it is pleasing, even beautiful. If there is anything complex or difficult about it, then it's not good enough. It is not the best solution. The best is always simple and lovely. Every genuine thing in this world is beautiful and harmonious. If it appears beautiful to you, it must be good, because simplicity, beauty and harmony are the acid tests for the right stuff. Marc Chagall, a painter with an inventive imagination, once said, "When I am finishing a picture, I hold a God-made object up to it—a rock, a flower, the branch of a tree or my hand—as kind of a final test. If the painting stands up besides a thing man cannot make, the painting is authentic. If there's a clash between the two, it is a bad art."

Appendix: Making Inventions 349

Producing a new and useful invention requires two essential qualities: knowing the tools of your trade and keeping an open mind. The first is relatively easy to acquire. You just go to a good school, study for many years, then work with smart people, learn from them, keep your eyes open and maintain your professional level. As simple as that. Knowing your trade is all that you need to become a good scientist, engineer, or manager, and it is often enough for securing a comparatively tranquil and comfortable lifestyle. That may be quite satisfying for most people. Not everyone is a born inventor and indeed it is a very good thing that Mother Nature keeps it that way. If every hen lay golden eggs, how would we make an omelet?

Alas, knowing too much about the field in which you are trying to invent something, can be dangerous. All too often, people who consider themselves experts feel that they know the full limit of possibilities and never attempt to go beyond that limit. Sometimes profound knowledge can creates a mental block. King Gillette spent eight frustrating years striving to invent and introduce his safety razor. Later, he said, "If I had been technically trained, I would have quit."

A little ignorance (or perhaps even a lot) may be of a far greater importance than knowledge. If you know all the ways things are done, will you have the imagination and strength to look further? Will you be able to put aside accepted rules and established practices? It is going beyond all this that usually makes an invention possible. What is most important is knowledge of a general kind and an understanding of the fundamentals of other fields of science and technology, for cross-fertilization may be quite fruitful.

Most of the inventions that I have made have been in fields other than the main line of my profession, medical instrumentation. In the beginning, I was somewhat ignorant in those areas and did not know that my solutions would be considered by the "experts" either impractical, or too simplified, or even unrealistic. When I invented my motion detector, I was very proud of its simplicity, low cost, and efficiency. I had no idea at the time that a similar idea had been patented well over forty years earlier and had since been forgotten. I simply reinvented it in a somewhat better and more modern form.

Another example. When I got the idea to devise an infrared thermometer for measuring body temperature, I knew very little about similar instruments. Of course, I could guess at how they might be designed and built, but never before had I seen such a thermometer, nor did I know the "right" way for doing such things. My knowledge was limited to the basic physical principles of heat transfer and the properties of the material I planned to use to make the sensor. It all seemed so easy and straightforward, to me that I went ahead and started up a company and began energetically raising money. I thought it might be a matter of merely a year or two to resolve some "minor" difficulties. Would I still have done it if I had known beforehand about all the hurdles and reefs on my way? Probably so, but it is so much better to be ignorant and persistent!

Finding the Problem

> There is no reason anyone would want a computer in their home
>
> -*Ken Olson*, president, chairman, and founder of Digital Equipment Corp., 1977

One serious challenge is finding a problem you want to solve. Not that such problems are in short supply. Quite the contrary, there are plenty of them, but the one on which you are going to bet your money, and often a large chunk of your life, must meet two essential requirements. The first is that the problem must have a practical solution within your lifetime. Usually, however, this does not become clear until you have already spent a great deal of time and money. The second requirement is that it must be *you* who finds the best possible solution, and you must do it *sooner* than your rivals. If the problem is solvable and nobody before you has gotten it right to the point where it becomes attractive to others, then you have a chance.

It is often said that an inventor is so enamored of his idea that he can fail to see its limitations or tends to overestimate its value or usefulness. Certainly, there is a lot of truth in that. An inventor can be preoccupied with his discovery and become quite insensitive to outside criticism, or even to the realities of the world. Fortunately, I have always had a sober view of my own inventions and have been able to sift out those which I felt had lesser potential. (Someone once said that if you want to have a sober view of things, divide everything you see in half.) You might call this marketing insight. Maybe it is. Thus, another important quality that an inventor should possess is a marketing mind.

Some people do marketing research in the hope of eliminating the risk associated with newness. That, however, is a waste of time and money, for there is no way to eliminate such risk, or even to reduce it. Market research simply does not work—you cannot do market research on something that does not yet exist. You can try doing opinion research, but that can be more of a hindrance than a help. The best guide for going for it is your gut feeling and intuitive insight. If you lack the talent for that, you may end up shooting yourself in the foot.

I used to know an inventor who had an idea for solving the problem of city graffiti. There is no question that this is indeed a problem in metropolitan areas throughout the world. Buildings and other civic structures with large surfaces are covered by ugly (or sometimes even pretty) pictures with a strong third-world flavor. This inventor approached this problem as a truly creative person. He did not propose setting up invisible fences with alarms around every wall in the city, or putting more police on the streets, as someone else might do. Instead, he suggested covering the walls of all the buildings in graffiti-infested areas with ... wallpaper. Every single wall, bridge, or underpass was to be covered with a removable waterproof wallpaper, which, of course, could be prefabricated to imitate granite, bricks, wood, etc. His idea was to provide the graffiti "artists" with sufficient surfaces for expressing their creative impulses and, at the same time, to reduce cost of restoration. The "artists" would cover the wallpapers with whatever graffiti they liked, then, once the surface was full, special city servicemen would peel off the used wallpaper and put up a fresh layer for the next round of "creativity." This inventor was so preoccupied with his idea that he failed to see, or did not want to see, several obvious drawbacks to his invention. The most serious objection was that the real problem with graffiti is not that it is too expensive to remove them and restore the walls to their original condition. The real issue is that city inhabitants simply do not want them on the walls at all, with or without wallpaper. People want clean streets and clean walls. Period. On the other hand, the graffiti artists would strongly oppose the wallpaper as well. They want their paintings to stay, not to be peeled off at any time. Undoubtedly, they would first peel off the wallpaper themselves and then paint on the real surface. The problem was genuine, but the solution was futile. This "invention" was clearly a wrong solution from any standpoint. The inventor was unable to see beyond his basic concept and so wasted a great deal of effort, time and money, all of which could have been put to better use.

Sometimes the opposite happens, when even an experienced inventor may not recognize an opportunity and gives up on something that might have great potential in a marketplace. Here are two examples. One of the world most celebrated inventors, Thomas Edison, strongly, if not violently, opposed the alternate current, which has now become the only practical way to channel electrical energy. I believe, though, that this was an ego issue rather than an actual misjudgment. Another great man, Chester Carlson, in 1937 invented xerography. His creative vision and high professional skills were unquestionable. Besides xerography, he had an idea for a much simpler product: a ball-point pen, but he dropped his work on the concept because he thought the pen would never work. The pen that worked was invented in 1938 by the Hungarian journalist Laszlo Biro, who saw quick-drying ink being applied by roller in a paint shop. He got the idea of the pen and developed it a few years later in Argentina where he escaped from the Nazis.

Types of Inventions

> High heels were invented by a woman who had been kissed on the forehead.
>
> -Christopher Morley

Generally speaking, an invention can be either technology driven or market driven. In other words, you can invent by attacking a problem from two different directions. One approach is to take some interesting device, material, or whatever, and think, what can I do with this that has not been done before? Is there anything new hidden inside this little old gizmo? Can it be adapted to do something unexpected but useful? I call this method the back-door approach. With it, you already have a solution for an as yet non-existent problem. What you need, then, is to find the right application for what you have. Often, you may even have to create a need for your solution. It is a marketing challenge that can pay great dividends if you find the right use for your invention.

Here is an example of a back-door approach. In the early eighties, I came across an interesting polymer film, metallized on both sides. It was a so-called piezo-electric film. That is, if you apply an electric voltage to its metallized surfaces, the film will flex. On the contrary, if you stretch the film, the electric voltage appears across its metallized sides, and you can measure it with a conventional voltmeter. I was fascinated with this interesting material. It had some electro-mechanical properties, but for what use? Of course, I knew what the piezo-electric effect was and how it can be used in various products. But all previous piezo-electric devices had been made of small rigid crystals. Here, however, we had a thin and flexible sheet of a plastic film. I had no idea what to do with it and began thinking about its possible applications. Initially, I knew of no product in which the film could be used. I simply had an interesting material in my hands. The film's manufacturer suggested using it to make microphones and loudspeakers. It was clear, however, that the material could be used for many more other applications—but for what? Eventually, I came up with several inventions using that film, because I figured out how some existing products could be improved by employing the film instead of the conventional crystals. In other words, I did not have any problem to solve, but I had a solution. I looked at the limitations in conventional products and was able to improve those products with the use of the piezo-electric film. My technology-driven inventions were based on finding the need, not the solutions.

The other way of inventing is more logical. It is the market-driven approach, where you know beforehand what you need to improve or what problem you are trying to solve. Necessity is the mother of invention. You set the goal, identify the problem and search for the solution. This way may be more direct, but sometimes it is too logical for producing a true invention. Nevertheless, by logically analyzing your task, you can generate plenty of interesting ideas. Sometimes these ideas go far beyond the initial task. You may start by solving a very narrow problem, but there is a chance that it may later grow into something bigger.

The great American inventor Charles F. Kettering had a market-defined task. In 1911, driving automobiles was very difficult for women because of the necessity of cranking the engine to start it. Kettering invented the electric "self-starter," which went on to become an essential part of any car or truck, regardless of the driver's gender. Here is another example. The summer of 1902 in New York City was hot and humid. A Brooklyn printer was having trouble with his color printing because the weather was causing the paper to change size sufficiently to result in distortions in the printing. He asked a young engineer, Willis Haviland Carrier, to solve the problem. Carrier found that air retained less moisture at lower temperatures. He designed a machine that blew air over chilled pipes and stabilized the amount of moisture near the paper in the printing press. The printing improved, and the solution, which had been developed for a very specific, narrow technical need, became the basis of home air conditioners.

Appendix: *Making Inventions*   355

Inventing Process

> Everything should be made as simple as possible.
> But not simpler.
> - Albert Einstein

Since inventing is a kind of mysterious act of creation, does that mean we are not in a position to understand how it is done? Can a recipe be developed for inventing? Are there any rules for finding a nontrivial solution? I have heard about many attempts to formalize the process, but nothing I have been seen is good enough. When the ambitious goal of reducing a creative process to a number of rules is achieved, inventing can be left to computers. But so far no one has invented an artificial inventor.

However, certain steps are known which may help you to keep moving in the right direction. Inventing is a trade, and like any trade it can be taught and learned to a certain degree. If you want to be a composer, it is very helpful to know music notation and learn to play at least one musical instrument. I know of only one very successful composer who got away without being able to read or write music. That was Irving Berlin. He was an exception, of course. Knowing your trade is not enough to make you a composer or inventor, but it definitely helps. So an inventor must be a learned person, preferably in a variety of fields. Also, the work of an inventor may be more successful if he learns how to organize his mental activity[1].

I will try here to outline a few essential steps for inventing a market-driven product. These steps may be helpful to get you moving in the right direction, for going in the right di-

---

[1] I am not a male-chauvinist. When I say *he* or *man* I actually mean either a man or a woman. When the American Constitution says *"all Men are created equal,"* I see it just as a historical curiosity of the English language. At that time, the word *man* meant person: that very same year, 1776, the Scottish philosopher and historian David Hume referred in his work to "all men, both male and female." In other countries a linguistic "sexism" is not a problem. In Russian, French, Italian, German, and many other languages the word *inventor* is masculine in gender, even if it refers to a woman. Hence, dear feminists, when I say man, I also mean woman. When I say he, I also mean she.

rection is itself already a great achievement. By organizing your mental processes, you can better understand the problem and come closer to the right stuff.

Quite often, when I try to solve something difficult and have just a vague idea about the problem, I start by writing a specification for the patent application. It sounds silly, doesn't it? Why write a patent for something that you have not yet invented? The reason for this is to organize your mental process, sort out your thoughts, and clearly define what is known and what is not, what your goal is, and what the logical steps might be for finding the solution.

The format of a formal patent specification is very clever. You start out by identifying the *field* of your invention. What is it that you are trying to invent? A home appliance, a new plastic, a toy, an automotive gadget, a musical instrument? The U.S. patent system lists numerous classes of inventions ranging from Abrading to Animal Husbandry to Weaving. At this point, of course, you do not need to worry about finding the legally appropriate class, just identify for your own benefit the most applicable area.

After the field or class is selected, you describe the *background*, that is, what is already known and why you think this is not good enough. By formulating the limitations of the previously existing art, you virtually set up your own goal. If there are no difficulties or deficiencies with what is already known, then why bother improving it? If you cannot write at least that portion of the patent application, then you do not understand your problem. This is a key point—clearly stating the problem and understanding what you want to accomplish. If you can do this, the job is already half done.

Once you have clearly stated your goal, the next step is to find a solution. It is conceivable to define a set of standard steps for inventing a new product. This set would consist of a number of rules for developing a new product or process or overcoming limitations in an existing product or process. The solution you find may be of an ordinary nature, or, to use the terminology of patent law, it may be something that might be produced by "those skilled in the art." That is, a

Appendix: *Making Inventions* 357

solution to the problem may simply be a result of taking logical steps and applying the standard rules of your trade. For example, if you want to pick up a low-level sound, the use a sensitive microphone is a known solution. If you want to pick up sound from only one direction, you may use a parabolic reflector with increased directional sensitivity. That also would be a straightforward step. Your design may be a great engineering achievement, but it is still not an invention. An invention, as we know, is something which is not totally logical, something which cannot be found in the textbooks. However, a good inventing textbook (though I have never seen one) can point you in the right direction, so that your creative mind will waste less time analyzing dead-end approaches. To carry our example further, if you make the parabolic reflector inflatable, like a balloon of a precisely defined shape, that could be an invention which solves some problem, such as the ease of transportation of the reflector.

A logical way of inventing is to dissect your problem, analyze it, and, eventually, get to the heart of the matter. You should keep asking various questions, even those which sound silly or irrelevant. In inventing, anything goes. Don't be embarrassed to ask the most ridiculous questions, something totally off the wall or completely crazy. By doing so, you may stumble upon an unexpected association which can ultimately lead you to your goal.

There are a few computer programs (one of them is IdeaFisher™) which can help point your imagination in the right direction. Such programs were designed for brainstorming and may be helpful in inventing process, though in a limited way. These programs contain large databases of words and ideas which can be searched and cross-referenced, producing those unexpected associations that can suggest avenues you might otherwise overlook. They are good only for prodding your memory. Obviously, they cannot provide the solution. That is still the job of a creative person and these programs are merely tools of the trade.

Just for the fun of it, let us invent something using the associative search method, like the one offered by some computer programs. Say we are looking for a new way to remove the cork from a wine bottle.

In the U.S. patent system a bottle opener might belong to Class 215, "Bottles and Jars". Our as-yet-to-be-invented corkscrew must do the job neatly and cleanly. We are frustrated with the conventional tools which can cause the cork to break, leaving cork crumbs in the wine and sometimes spilling the wine, and which require excessive force to remove the cork. These are the limitations of the "prior art" which we will try to remedy.

The process of searching for the most promising solution begins with looking for associations between your starting point (which in our case may be either a cork or a bottle) and anything which might have even a very remote connection with it. The associative search can lead us to a string of words, all ultimately related to the initial word *bottle* and each linked closely to its neighbors:

For example: 1) ***bottle***—*glass—liquid—evaporation—pressure—temperature*, etc. Each word must have a strong association with the words directly before and after it, forming a kind of chain. We now have our list, but it looks as though these words are irrelevant to the purpose of our search. However, if you think carefully about each word, applying your background knowledge and imagination to it, you will start to see the profound meanings in these associations. So your next step will be to think about every word in the chain.

Our evaluation of the associated words from the list we have compiled begins with the second word in the series, *glass*. We already know that a bottle is made of glass, which doesn't help us. Next we have:

Liquid—Yes, there is liquid in the bottle. So what? Keep going.

Evaporation—Wine evaporates inside the bottle, but so far this doesn't give us a clue.

Pressure—Beneath the cork inside the bottle a mixture of air and wine vapors produces pressure. What do we know about this pressure? Obviously, the pressure is nearly equal to the outside atmospheric pressure, otherwise the cork would not stay in place. An excess of pressure on one side or the other would push the cork either in or out.

Wait a minute! We don't want the cork to stay in place. What we want is to get the cork out of the bottle. Do you see? Things are starting to cook! The word *pressure* seems to be telling us something. Let's go on and look more carefully at that word, because now we want to find a simple way to change that pressure. This time we use *pressure* as an initial seed word to assemble a second set of associations.

2)***pressure***—*temperature—volume—flow—pipe—tube-plumbing—pump—piston*, etc. Again, we start with the first word in line. *Temperature* is something we do not want to change (no one wants hot wine, unless it is grog or Japanese sake). None of the next several words suggests anything we can use. At least, I do not see it, though another inventor might well view them differently. That is what makes every inventor unique.

One of the associative words is *piston*. Does a cork look like a piston? Can it act like a piston, as in a car engine cylinder? Here we have something interesting. In order to move the cork out of the bottle, we want to increase the pressure inside the bottle and the cork, like a *piston*, will move upward along the bottle neck until it pops out. But how do we manage this?

Let's go back to our second string of words. Perhaps there is some clue there.

Volume—We see no way to do anything with volume. It is fixed for the time being, though that will change when the cork comes out.

The next is *flow*. The wine does not flow as long as the cork is in the bottle. Anything else? The vapor inside does not flow either. Air outside the bottle may flow, but how? I do not know... What's next?

Pipe—So far we have no pipe, but if anything flows, it can flow inside a pipe. Can we make the wine flow through a pipe? Yes, we can put some kind of pipe or straw into the bottle and make the wine flow through it. But first, we need to remove the cork—which was our prime goal.

Plumbing—Again, we could do something with several tubes, or pipes. We could arrange them into a plumbing network, but how can we do this while the cork is still in place, and to what purpose?

The next word is *pump*. With the cork still in place, we could pump wine out of the bottle by means of a tiny tube inserted through the cork. To do that we would have to increase the pressure inside the bottle or decrease it outside. That might work, but pumping wine through a tiny tube would be too slow and hardly elegant. Besides, our goal was to remove not the wine, but the cork. So now we're back to looking for a way to increase the pressure.

Tube... A small tube... A tiny tube, like a *hypodermic needle.* This sounds very interesting. We go back to our search and look at the words again, linking together the things that seemed promising: pressure, piston, pump, tube, hypodermic needle. Aha! We are getting something interesting here! Why not do what we have already found—insert a *hypodermic needle* through the cork into the bottle, and rather than pumping the wine out, instead *pump* air from outside into the bottle. The *pressure* inside will increase until it overcomes the forces of friction, at which point the cork will slide out, just like a *piston*. To make a practical device, what we need is a hypodermic needle and a small air pump, and our invention is complete.

The rest is straightforward. It is no big deal to design an air pump, similar to that used for bike tires, only much smaller, of course, and attach a hypodermic needle to the end of it. Actually, at this point the research and development starts. To make a reliable product, a designer must determine how the invention will work under various circumstances. For instance, what is the range of force required to push the cork out? The force may depend on the type of cork, its size, the surface of the glass, the chemical composition of the wine, its age, fermentation, etc. Also, the optimum dimensions of the hypodermic needle must be determined, so as not to fracture the cork and to make the pumping easier. When the greatest possible pressure is determined, the air pump will be designed to assure both technical performance and ergonomics. Naturally, at some point, an industrial designer will be engaged to find the most convenient, aesthetically appropriate, yet technically acceptable shape, color, materials, etc. This is the typical process of product research, development, and design.

Often, a designer may discover that a product does not work as planned. In this case the inventor must go back to the drawing board to develop additional improvements. For the product designers, it is almost essential to work with the inventor up to the very last phase, when the product is actually ready to be shipped to customers. Quite often, the inventor's participation never ceases. I developed the fundamentals of my infrared ear thermometer during 1982 and 1983, the patent application with its detailed description was filed in early 1985, but even today, in 1994, when I am writing this book, I often take part in resolving various questions related to its production and quality assurance, or consulting customers.

While product design is an essential step in the making of a practical device or process, it differs dramatically from invention. As I have already noted, an invention is something unknown, not naturally and directly evolving from previous experience. All previously known corkscrews were gripping devices which physically engaged with the cork in order to pull it out of the bottleneck. Creating gas pressure inside the bottle was something that was not obvious, and that is what makes it an invention. On the other hand, the design of a small air pump is an engineering project. Yes, the pump must be made to a smaller scale and must probably be operable by a thumb or index finger, but it is still the same old air pump doing basically the same job, simply modified somewhat. Its design is fairly obvious, yet it requires the skills and knowledge of the trade.

In many other cases, the border line between design and invention is not as clear or self-evident. Often, this line is illusive and depends greatly on interpretation. That is why it

Appendix: Making Inventions

is so difficult to argue with a Patent Office examiner, who often says, "This is obvious!" It may be obvious to him, though not to many others. But that is one of the difficulties an inventor must overcome—arguing with the patent examiner. For this, you need a good patent agent or patent lawyer. But more about that later.

Coming back to our corkscrew, I just want to mention that we have no need to build a prototype or design anything, because you can go to a store and buy one just like the one we have "invented" here. A corkscrew with an air pump was conceived in Switzerland and has already been "reduced to practice," i.e., designed, fabricated, and marketed, and enjoys a broad popularity.

Of course, I do not know if the inventor of the pneumatic corkscrew followed the same path we have just described. Quite possibly, he went about it in a less logical and a more intuitive way. An inventor is a curious person, who sees things that others do not. Perhaps one night he was having a good time at some party where he noticed a bottle of Champagne being opened. The sparkling wine pushed the cork with such force that it shot to the ceiling. Perhaps this triggered his vivid imagination and he thought, Why not do the same thing with a non-sparkling wine? If there is no excess of pressure inside the bottle from natural fermentation, let's cause that pressure to rise artificially. And to increase pressure (as every child knows) you need a pump. Or, it may be that it was something else altogether. Who knows how a creative mind finds its way, though all roads lead to Rome.

Patenting

> Necessity is the mother of invention,
> but patent rights is the father.
> -*Josh Billings*

Okay, let's assume that you have found the right solution. Now, it's to be or not to be, that is the question! Alas, for the entrepreneurial inventor, that is not the question at all. Obviously the answer is *to be*.

The next logical step is to protect your idea. Of course, you have kept your dated notebooks, even had their pages notarized. Perhaps someone has already told you that the cheapest way to secure your distinction of being first is to make a full written description of your idea, seal it in an envelope and send it to yourself by registered mail. When it arrives, do not open the envelope and keep it in a safe place. I did things like that many times, though later I learned that such protection is worth little or nothing. In the case of a dispute, no court will seriously consider those sealed envelopes, unless you kept constantly working on your invention after that. You must show diligence in order to claim that you did something sooner than someone else. A conceived idea which has been sealed in an envelope and then left alone is considered abandoned and the inventor loses his rights to it. There is only one real way to protect an invention—to file a patent application with the Patent Office. Unfortunately, it is an expensive way to go about it, but unless you are making your invention just for the fun of it, you had better be prepared to spend the money.

It makes sense to apply for a patent only if your idea meets two requirements. Number one is that the patent *can* be obtained. Number two is that you *need* the patent. Before spending money to acquire a patent, you should ask yourself why you need one. Apart from the question of ego, a patent may be necessary in order to create a monopoly for yourself from which you will then benefit. It is funny, but the patent system, which is set up in the Constitution, is the only monopoly allowed in this country, where anti-trust laws are so vigorously enforced (the anti-trust law was not provided for in the Constitution). Yet that kind of monopoly is a healthy and fully justifiable reward and incentive to an inventor for his pioneering work.

If you have a good patent attorney, you can almost certainly get a patent for almost anything. Even if your idea is not entirely new, a cunning lawyer can twist the disclosure and the claims in such a manner that the patent examiner will have no choice but to allow the patent. Usually, this can be done by narrowing the claims. It is very difficult to acquire a patent which protects everything. The Patent Office does not like to grant broad patents. But the narrower the claims the easier it is to get a patent. Of course, the value of a patent whose claims are so narrow that they protect virtually nothing may be very questionable, yet even very narrow claims have their own mysterious value. This value is of a psychological nature. Competitors do not like to have someone else's patents intrude into their fields, even if such patents are very narrow. In effect, every patent, broad or narrow, works as a sort of scarecrow. And in many cases this is reason enough to have one. It may slow down the competition and buy you some time.

On the other hand, getting a patent too soon, before your idea is mature, may offer a hint to potential competitors who can then gain the upper hand by learning how you did it and using this information to get around your solution. I myself have several patents which were suggested by new patents issued to my competitors. From these patents, I learned how they did it, found some deficiencies in those patents and filed for my own "get-around" patents, which became much stronger and virtually pushed those competitors aside.

If you are unsure about the strength of your patent, an alternative strategy is to file the application, but use every means to slow the issuance of the patent as long as you can. Your patent lawyer will know a number of tricks for managing this (actually, he will love that tactic, for he will be able to charge you for the extra time). After filing for a patent you can start selling your product with the statement "Patent Pending." (In fact, you can start selling even a year before filing for a patent.) Keeping the patent pending longer can buy you a great deal of time, because what exactly is pending is a secret until the patent is issued. Unfortunately, that will work only in the United States, where the filed applications are kept confidential until the time the patent is issued. In many other countries, for instance, Japan, the application is published immediately, so that the whole world knows what you did and how you did it. And that may not be to your benefit.

Sometimes, it's wiser not to file a patent application, but to keep your idea as a *trade secret*. One company I worked for had developed a unique process for the fabrication of miniature plastic components. Initially, the management of the company wanted to patent the process for making such parts, but after thinking it through, they realized that the patent would disclose the most vital details of the process. If, somewhere else in the world, someone decided to duplicate it, he could become a very serious competitor with almost no risk of prosecution, since the owner of the patent would have no means of policing the infringement. Just by looking at the plastic component, it was impossible to tell what kind of process had been employed to produce it. The company, of course, could never get its inspectors into every factory in the world to investigate how parts were being manufactured. Keeping it a trade secret was much more secure. Someone out there might well reinvent your tricks, but in the meantime you keep the benefit of being first.

So sometimes it is good to have a patent and sometimes it is not so good. However, not getting a patent usually makes sense for a company with significant assets, which can afford to produce and sell products and keep its trade secrets secure.

But what about an individual inventor? Can he afford to keep trade secrets and not file for a patent? I do not think so. Trade secrets simply have no value to an individual. And that is another good reason for an entrepreneurial inventor to obtain a patent. Let's say you want to sell the license to your invention, or you want to raise money from investors to start an enterprise. When you have a patent, or at least a patent pending, you own a certain intellectual property. If you have no patent and none pending, the invention quite simply is not yours in any legal sense. You cannot sell what does not belong to you. Or if you try, you must disclose the minute details of your invention and run the risk of losing it.

Those glorious days when many companies would agreed to sign documents protecting inventors (those documents are called *disclosure agreements*) are gone forever. Now the situation is just the opposite. A company you approach with your invention will make *you* sign a disclosure agreement stating that you are disclosing everything at your own risk and that the company is under no obligation to you. This agreement protects the company, not you. Hence, if you have not already secured a filing date or, better yet, acquired a patent, either a reputable company will not talk to you, or you yourself will not be at liberty to talking to them. Raising money or selling something which is not protected by a patent is like beating a dead horse. If you want to sell it, you had better patent it.

Unfortunately, there are other, more sinister reasons for obtaining a patent. I have seen how a sleazy inventor once sued a corporation for "infringing" on his patent. There was really nothing to it. It was a typical case of legal harassment. The patent claims were so narrow and irrelevant that the company would have had no problem defending itself. But even though winning the trial was a sure bet, the company decided not to go for it and to settle out of court. Do you know why? Simple accounting: the legal expenses would have been several times higher than a settlement with the "inventor." In effect, the company was afraid not of the cunning plaintiff, but of its own lawyers, who would miss no opportunity to suck blood from their client. Anyway, that extortionist owned a patent which protected virtually nothing, and yet got for it a pretty fat chunk of money. That is the reality and the cost of being in business in a country with a litigation-happy climate.

Certain other con-artists, who make their living in the shadow of creativity, make patenting a profitable business. Once I met N., a rich man who bought ideas and patents from small inventors who could not afford or did not know how to proceed further. He had several hundred patents in his own name and many others that had been assigned to him. He and his agents would scout about for the big guys, that is, for companies who had a lot of cash and had apparently similar products on the market. He then sued them for "infringing" one or the other of his patents. Some companies settled with him, just as I have described above, while others were forced to buy the patents in question from him for large sums of money, but most of the cases he lost. In fact, he lost nearly 90% of his litigations, but the remaining 10% was more than enough to secure a very comfortable lifestyle for himself and his lawyer-partner. It was repelling to talk to that "gold-digger." He was a parasite on the body of human creativity. I hate that type of business, but it is a reality of American life and you either take it or leave it. So far everyone just takes it, as far as I know.

Fortunately, most successful inventors make money the old-fashioned way. They earn it by making their inventions work, or, to use the legal jargon, "by reducing them to practice."

If it is your first invention, what you need is to find a good patent agent. And I do mean an agent, not a lawyer. Agents charge far less than lawyers but can do most of the same jobs, and often much better. Indeed, many patent law firms, when overloaded with work, will often hire these self-employed patent agents to write the patent applications and do all difficult ground work on the patent: prior art analysis, filings, and even prosecution. Naturally, the law firms charge the clients their own high rates, then pocket the greater portion of the money and pay the rest to the agent. Essentially, a law firm acts as a middleman. Who needs a middleman when money is tight and a better use can surely be found for it? In the beginning, I did not know all these mechanics and was forced to dilute a great deal of the interest in my inventions to raise money to pay high legal fees.

Once, when I was living in New Haven, I had an idea for an invention for which I wanted to file an application with the Patent Office. At that time, I was virtually broke and could not afford the one lawyer I already knew. His fee was $150 an hour, and who counted those hours? There is a joke about a dead lawyer standing before the Pearly Gates. The registering angel declares that he is proud to admit the first person older than Methuselah himself. The lawyer protests, saying that he is only fifty-five. The angel replies, "No, according to your business log book, you are one thousand two hundred and fifty years old."

I opened the Yellow Pages and there, listed under *Patents*, I found the name of Bob Seemann, a patent agent. Not knowing what that title really meant, I called him and he said that he was a registered agent (it is very important for an agent to be registered with the Patent Office!). He said that as an agent he could file and fully prosecute any patent application. The only thing he could not do in the line of his profession was conduct litigation. I met with Bob and asked him to show me some patents he had obtained. I asked him about his area of technical expertise, where he had studied, what law firms he worked for, whether he ever went to Washington to talk to patent examiners, and how long it would take him to write a disclosure. And most important, I asked him to give me his complete fee schedule, which described in details his services and fees. His basic charge at that time was $60 an hour, which was quite modest compared with the lawyers I had recently talked to.

That was about eight years ago. Since then, Bob has worked on many of my patents, never losing a single one of them, and rescuing two which had been nearly abandoned by other lawyers. When you need a patent, it is very important to find such an experienced and reliable person. He can be a real asset to your inventing business.

If, God forbid, you must go through litigation, a patent agent will not do. You will need to find a patent attorney with a good track record in winning similar cases, which is, of course, an expensive but often a necessary expenditure. However, for most practical and undisputed cases, a good patent agent is all you need.

When an inventor gets an idea for something he wants to patent, he hires a patent agent or attorney. Often, the only thing he brings to his legal representative is a model or prototype, or just a piece of paper with a very brief description of the idea. A good agent or lawyer can write quite an impressive patent specification with even that slight background information on the product or process. Obviously, this is not at all a wise way to obtain a patent. Even the most brilliant patent lawyer is not an inventor. Most likely he is not terribly familiar with the subject matter. It will come as no surprise, then, that the specification he writes may contain a number of errors and the claims he constructs may miss many valuable options and versions of the invention. This can result in a weak patent, which means that the inventor's legal protection will be very limited. I have learned from experience that the best way to work with an agent or lawyer is to come up with as detailed description as possible beforehand. Try to put in the body of the specification every possible option, alternative, and preference for the implementation of the idea. Try to see whether it is possible to change any critical components and still have your invention work. For instance, you may need a tubular pipe. Is it important that it has a circular cross-section? Can it be triangular, square, etc.? If you state in the specification that it is round and the claims call for a circular cross-section, it will be very easy to get around your patent simply by changing the profile of the pipe. Your goal is not just to obtain a patent. It is important to create a "road block" against those who may want to get around your invention. By doing your homework before taking your idea to a legal representative you will be doing yourself a tremendous favor.

It is best if the paperwork you bring to the patent attorney or agent is prepared in a format similar to a finished patent. People in the legal trade are accustomed to reading such documents and will be much better able to understand your idea quickly and translate what you have written into what is required for filing with the patent office. Naturally, this will save you a lot of time and money. The outline of your groundwork should include: 1) the field of invention, 2) the background, that is, what is already known and how things are done at present (include any patents, references, sales literature, etc. of such previously known devices or processes), 3) the limitations and deficiencies of the prior art which you hope to reduce or eliminate, 4) a condensed description of your idea, stressing what you believe to be its most innovative component, 5) a detailed description of your idea with an example of its practical implementation. Do not supply a set of claims. Your lawyer or agent knows better how to construct them to provide the best legal protection, but make sure that you understand what he writes and do not miss anything important.

Whatever you try to patent must be "reducible to practice"—it cannot be simply an idea which has no technical solution. Everything you put down must have a way of being made today. You cannot just hope that something will be discovered in the future. If, for example, you try to patent a device for ESP communication, you had better be in a position to prove without doubt that such extrasensory perception in fact exists, before you use it in your device. Everything you do must be in compliance with the laws of Nature as present-day science defines them. Do not try to patent a machine for producing perpetual motion without using an energy source—it is against the laws of science.

Promotion

> Diligence is the mother of good luck.
> - Benjamin Franklin

> Drill for oil? You mean drill into ground to try and find oil? You're crazy.
> -drillers to Edwin L. Drake who tried to enlist them to his project to drill for oil in 1859

Now, let's talk about turning your invention into a cash inflow. If you are a poet, selling your verses, while not easy, is a fairly straightforward process. You need considerable patience, plenty of postage stamps, and perhaps a good literary agent. However, if what you want to sell is not a poem, but an invention, is the process as straightforward? Will the world come to your doorstep if you invent a better mousetrap? No way. The world is fussy and whimsical. It behaves like a prima donna in a cheap opera. And that is the real difficulty with inventing. You, as an inventor, not only have to invent a machine or process and make it work, which is often a tremendous task in and of itself, but you also need to convince the world that it cannot live without your invention.

You have to find the right people who want to go along with you. You need to find a source of money to finance your venture. Entrepreneurship requires stubborn persistence and a zealous belief that what you are doing is right. It demands great skills in diplomacy and salesmanship in order to sell your idea to investors or marketers.

And you must not be greedy. You must be prepared to lose with an easy heart. A real inventor has the character of an adventurer and a risk taker. At the same time, it cannot be stressed too strongly that a good inventor-entrepreneur, while a risk taker, is not a risk seeker. Only fools look for danger in business. A serious entrepreneur must be very prudent, planning carefully, calculating wisely, and taking only those risk which are unavoidable.

To be an idea generator simply is not enough. The inventor must also be willing to fight for the survival of his idea. That is what makes the entrepreneurial inventor so different from people of other creative professions. Van Gogh never lived to see his *Sunflowers* sold for thirty millions dollars. Even if an inventor does not commit suicide as van Gogh did, life is still too short. Inventors cannot afford to wait, for the lifetime of a patent is even shorter than that of a human being—only 20 years from the day of filing in the United States, though in reality it may become obsolete much sooner. If you want to do it, do it now, or never.

There is an inevitable series of steps that must take place if you want to make money from your invention. These steps are: invention, protection, and commercialization. Promotion or commercializing your idea is an entrepreneurial business. A business is a business, no matter what you make or sell—soft drinks, wallpaper, railroad engines, or inventions. The inventing business is just like any other. It must be organized and managed as a small business, including all those aspects of a business which you may or may not be willing or able to undertake.

There are several ways of reducing an invention to practice. However, it should be clearly recognized that no matter what you do or how you do it, it is still a *business* venture. And like any business, it resembles a three-legged stool. One leg is your product or invention, another is the business structure and management (such as engineering, manufacturing, marketing, sales, public relations, etc.), and the third is the financing. Remove any one of these legs and you will fall. I have already talked about bringing about the first leg, which is the inventing. Now, I am going to discuss some possible ways to provide the other two legs. This book, however, is not about starting up and running a business, so I would suggest that the interested reader find a business textbook

Appendix: Making Inventions 371

and study it carefully. Here, I am sharing my personal experiences, which may serve as an illustration but are not a substitute for a good text. What follows is a non-exhaustive list of possible strategies for going into the inventing business and raising money for its operation, strategies which I either undertook myself or seriously considered.

Doing it yourself.

I believe that this is the most reliable way to make it happen. By starting your own business dedicated to capitalizing on your invention, you keep it under your own control and do not waste time on selling your idea to others. Time is often a crucial factor on the road to success. By jumping on something right away and starting it rolling, you will not only have the advantage of doing it sooner than the competition, but your example and desire to take chances may prove to be the most convincing argument to potential investors. If you need more money for further work (and most likely you will), equity investors may be the only source of money. However, if they come at the later stages of the venture, the dilution of your stocks will be smaller and you will be able to maintain more control in the future. Besides, finding such investors will be much easier, for very few investors are willing to consider a bare and unproved idea, but many will take a look at a up-and-running business with a cash flow in both directions.

Unfortunately, not every inventor is a born businessman. God usually chooses different heads into which to plant these talents. Fortunately, I have had no illusions about my business talents. I knew that I can be creative in various areas, but that I have no strong ability for running the day-to-day operations of a business. Certainly, I have some skills in that area, but my temperament and way of thinking often distract me from an essential business discipline. If an inventor does not feel confident about running a business himself, the solution is to find a partner to take care of that side of things. The best synergy is when one person sticks to the technical stuff while the other partner handles the business matters.

Doing it with a partner.

Finding a trusted, reliable, honest, hard-working, experienced, and talented business partner is extremely difficult, if not impossible. At least, I have never had much luck at it. I have approached various people with offers to join me in my ventures. Some were honest enough to tell me straight off that either they were not born entrepreneurs or they did not feel like taking such a chance, or just simply turned me down for some other reason. I appreciated and liked that directness. Some other people gladly accepted my offers (they thought that I was going to make them rich), but later on I had to pay dearly for making wrong choices.

When you consider someone, be very, very, very careful. Do not be fooled if that person tells you how great your invention is and that he sincerely believes that it is the best thing since peanut butter and jelly. This is not a matter of affection, love, or even sympathy. A partnership is a matter for cool calculation and background checks. The best business union is a marriage of convenience, not one of love. Another rule of thumb is never to go into business with a close friend or relative. By doing so, you run the serious risk of losing both the business and the friend.

Doing it yourself or with a partner, while the best strategy from the standpoint of maximizing your future profit, is also the riskiest. It may work and it may not. The chances are that you and your partner will be broke and need to start from scratch. Do you have the guts for that? Besides, not everyone is financially secure enough to quit a job, start a business and provide funds for its operation until other investors are found.

Even while believing that the "do-it-yourself" method could be the best way for me, I never did it myself, for I felt strong obligations to my family. I was new to this world of free enterprise, I had no rich relatives, no savings, no property, no one to borrow money from, and nothing to use as collateral. Like a newborn baby, I had come naked to this Free World. Perhaps if I had been alone and had any assets at all, I might have tried it. But my only valuable possessions (though quite useless as collateral) were my wife and two children, whom I had to support and whom I had no wish to put in jeopardy simply because I myself was a risk taker. That is why with all my inventions (until I became financially secure) I always took another path, which was to find financial backers in the very early stages of the venture.

Raising money before you start.

Business textbooks list several sources of money for an enterprise. Such traditional institutions as banks simply do not work for inventors or start-up companies. Of course, if you have equity in your house, you can always get a second mortgage, or you can use some other valuable possessions to secure your loans. But if you have no property that you are willing to risk, or cannot afford to do so, you need to find another source of money. Debt financing, that is, getting a loan, is a nearly impossible source of venture money. The reason for this is that hardly anyone will lend you cash for something intangible or highly speculative. Banks are not in the gambling business.

The most reliable method in terms of reducing the risk of personal financial failure is to use someone else's money, that is, equity financing. Unfortunately, security and success add up to a constant. The more financial security you hope to retain, the less your chances for financial success. I have known a number of inventors who refused to bear any risk at all: whether it be risking their own money or sharing future earnings with someone else. Every single one of them ended up a failure. A high risk anticipates a high reward. If you are afraid of taking a risk or cannot afford to do so (as I could not), your only option is to pass that risk on to someone else who is in a better position to gamble than you are. In exchange for their money, you give them part of your equity, or, in other words, they assume both a portion of your present risk and a portion of your future earnings. Such people are called start-up investors. By nature, they are risk-takers. Unfortunately, they are a very rare, indeed almost extinct, species. During my many years in America, I have met only a handful people who fit that category, and only one of them had the courage to stick with it all the way to success. That man risked a great deal of money on one of my inventions, but his reward was enormous. Sharing the risk, of course, means sharing any future profit, but that is the price of being in the inventing business.

Individual start-up investors are rare, but there are still some of them around. The hard part is finding them. Talk to your friends, acquaintances, and relatives. You may get a good tip from your accountant or a lawyer. It could even be worth your while talking to your bank branch manager. Though the bank itself will not invest, you may get some good leads to people with deep pockets and adventurous natures who may consider putting some money into your enterprise.

Try to stay away from relatives or friends. Instead, look for strangers with whom you can establish a strict give-and-take business relationship. Hardly anyone will consider investing millions, but if you can convince them of the benefits of your invention and present a valid business plan, some may consider putting up ten, fifty, or even a hundred thousand dollars. As a rule, an investor never puts a large chunk of money into a risky business all at once and then patiently wait for you to make him richer. He will feed you little by little, depending on your progress and needs, and in exchange, of course, for more and more shares. Such a practice is understandable, as any normal person wants to minimize his risk and financial exposure.

Another source of money is venture capital companies. This name is highly misleading, because these companies rarely have their own capital and, frankly, they have nothing to do with ventures. But that is what they call themselves and I have no choice but to use the same name. Such companies often manage money for large pension funds, investment groups, rich families and so on. The people who run the venture capital businesses never invest their own money. In most cases, they are just salaried employees. And as employees, the foremost concern of these people is to preserve their own positions. They are interested in job security, not in risking someone's money or making a big profit. One Canadian venture capitalist told me, "We would be very happy with a seven to twelve percent return on our investments and we want zero percent risk with that." That is not what I call a venture.

Very rarely, and only if you have an extremely hot invention which a number of analysts rate as potentially highly successful, will a venture capital company consider you. And if a company does invest in you, it will demand an arm and a leg in return. Venture capital companies are extremely hungry for equity. If they give you money, be prepared to become a minority shareholder and lose all control over your enterprise. These companies, however, may be more receptive to your needs if you already have an up-and-running business of manufacturing and sales. They may come to you with their money and bring in their own management skills and business insight. But at the early stages of your venture, do not count on them.

You, as an entrepreneur, must be aware of the value of your business. The more progress you make and the more success you can demonstrate, the higher the value of your stock. For instance, when all you have is a basic idea, the price for stock in your venture may be just 5¢ a share. When a prototype has been built and tested the price may jump to 25¢. And once a patent has been granted, the price may grow to 50¢, and so on. On the other hand, the money that is being invested is real money ("Money is King," an old proverb says), while you are still selling a dream. Until the dream becomes a tangible reality, its value is questionable. In other words, an inventor should be ready to sell the initial stocks cheaply, but not too cheaply—when an inventor's equity is diluted too much, that inventor loses his incentive and the entire enterprise may fail.

I believe that raising money from individuals is the most practical way of financing an invention. It was the way I got money for most of my own inventions. Yet it is a very, very difficult way to go. All too often, an inventor has to give up a great deal of his own interest in the invention, but unfortunately nobody has yet come up with anything more fair or reliable. Modern civilization has developed no suitable mechanisms for promoting innovations. Even the government offers no incentives whatsoever to those who invent or provide financing for inventions (such as tax brakes, for instance). As in the remote past, everything depends on luck and the inventor's persistence in finding a sponsor.

Another way of getting your venture financed is to reach an agreement with an existing business which may provide funds for your start-up operation. In return, that business might become the exclusive owner of your invention, or it might receive the right of first refusal when the device is ready for commercialization, or it might simply own stock in your company. Potentially, this can be the most attractive arrangement. It can be very good for both the inventor and the company. However, I have heard of only a handful of deals involving individual inventors and corporations which served in the capacity of financial backers. Who knows how many companies have missed out on great opportunities with the modest price tag of risking relatively little money?

In Chapter 13 I told the story of the Instant Thermometer, which was my most successful invention. It became successful because several essential factors came together. One was my strong belief in the product and my readiness to fight for it to the very end. The other was that I had the luck to find an investor with sufficient resources and a willingness to take chances. It was also important that I was able to struggle with that investor for my share in the success, otherwise his overwhelming desire to leave me with almost nothing could have force me to move on to another project and abandon the Instant Thermometer. I have said it before and repeat it here again. To succeed, an inventor must be a fighter. His enemies are technical difficulties, lack of money, lack of knowledge, and competition. In the inventor's quest for success, he may also end up struggling with his friends, his wife and children, his colleagues and even himself. Never giving up and fighting to the end—that is the essential condition for any achievement.

Strategy

Once you have started a business for the commercialization of your invention and secured financing for it, it is important to manage it as an innovative business. Defining your strategy is crucial. Do not think big, do not try to serve a global market. Try to focus on a smaller niche, answer a particular need. With every new technology there is considerable lead time and a long learning curve. If you focus your efforts and money, your chances of succeeding will rise dramatically. Later on you may well expand, but in the beginning it is too risky to think big.

Quite often, success lies not in the invention itself, but in the strategy for bringing it to the marketplace, for building a business around it, and for focusing that business not on the invention, but on serving a need in the market. That need may have existed before you entered the market or, even better, you may have created that need through a carefully planned marketing campaign. If you create a need which you then fill—you are holding fortune by the tail.

We all believe that Thomas Edison was the inventor of the light bulb, yet in fact his design was not the best. Far superior was the light bulb invented by the British physicist Joseph Swan. For quite some time Swan tried to find someone to take an interest in his invention. His only desire was to sell his patent and to collect the royalties. The person he found was Edison. Edison bought out Swan's patents and used them to produce his own light bulbs. In effect, Edison did much more than design the bulbs. He was concerned with installing and selling *electric power,* of which the light bulb was just one essential component. Edison built a great business and is considered the father of the electric light bulb, even if it was actually invented in various forms by a number of inventors ranging from the Russians Yablochkov and Lodygin to the Englishmen J. W. Draper and J. Swan and the Americans M. G. Farmer and Edison himself.

I am not going to talk here about the various strategies and methods for running an entrepreneurial business. First of all, I do not consider myself an expert in the field. Secondly, this book is not about that. The reader who is interested in that subject can find plenty of good texts, for instance Peter F. Drucker's *Innovation and Entrepreneurship.*

Selling your Invention.

> This "telephone" has too many shortcomings to be seriously considered as a mean of communication. The device is inherently of no value to us.
>
> -*Western Union internal memo*, 1876

Many inventors look for an easy way to cash in on a good idea. The typical thinking goes like this: "There are companies out there who are in a similar line of business, but their existing products are not as good as my invention. Instead of competing with them, why not sell them my idea, and both of us, I and the company, can live happily ever after. The company will produce and sell the product based on the invention and I'll keep collecting royalties for as long as it lasts, without any of the headaches of being in business for myself." Right?

Wrong. A company which has something similar on the market is not your ally, but as long as you do not seem to pose a threat, it is not your enemy either. Such a company simply does not care about you or your idea. If you try approaching existing businesses, you will face, at best, mere indifference, and at worst, and more probably, violent rejection. The reason for this is the infamous NIH syndrome I mentioned in chapter 9. Internal opposition within the company may be so strong that even those individuals in top management who might be in favor of it will not take a chance on it or risk going against the grain.

Of course, it is theoretically possible that there are companies out there who are dying for your invention and will meet you with a marching band and flowers the minute you arrive on their doorstep. This may be so, but I have never heard of such luck. I spent years trying to sell several of my inventions to already existing companies. I had dozens, if not hundreds, of very hopeful and promising meetings, not one of which ever materialized into a deal. I even signed several deals. In two cases, the companies later had no idea how to rid themselves of me. In the end, they did manage to get rid of me, but they lost in a long run, while I made it. That is the reason why I cannot place much faith in selling an invention to an existing business. Selling your business

Appendix: Making Inventions

to another business is fine, but selling just a patent is very, very difficult, and even if you manage to do pull it off, it may not work in a long run.

Let's assume that you have been lucky and found a company that wants your invention. What type of deal is best for you? Since this is not a manual or a textbook on the art of negotiating or licensing, I will just outline what I consider to be a fair deal. By fair I mean to both the inventor and the entity which obtains the license from him.

It is often very difficult to put a price tag on an invention. Some inventors may overestimate the value of what they have done, while others may be far too modest. In the early 1870s, Thomas Edison invented the printing telegraph. He took it to General Lefferts, the president of *The Gold and Stock Telegraph Company*. When asked what he considered a fair price for his invention, Edison's first impulse was to ask for $3,000, but he hesitated and said instead, "Make me an offer." He was quite surprised when the offer turned out to be $40,000! Steadying himself, the young inventor quietly said, "Yes, I think that would be fair." It is a lot of money even today, but 130 years ago it was a fortune.

Before an invention is completed, patented, and offered for sale, its author has spent an immeasurable amount of time, and often a great deal of money. The issue, therefore, is not what the inventor should get, but what the value of the invention is to the buying party.

In most cases, the license for a patent is sold on an exclusive basis, that is, only to a single organization. A non-exclusive license can be sold to a large number of different groups or companies, which makes sense for them only if they are not concerned about competing with one another. For instance, an inventor might offer a bottle cap for sale on a non-exclusive basis. Companies in competition with each other, for instance Coca-Cola and Pepsi-Cola, while rivals, would not necessarily compete for the bottle caps. They might both be interested in buying a non-exclusive license for the same cap.

However, when the license is exclusive, the inventor should ask for a so-called *licensing fee* up front, which is in addition to the royalties. This fee can be considered as payment for the exclusivity and is intended to compensate the inventor for the increased risk of doing business with only one buyer. Depending on the value of the invention, the total size of the deal, and the strength of the patent, the licensing fee can vary widely, anywhere from five thousand to half a million dollars.

The other portion of the compensation is the royalties, which for patents range between 2 and 7% of the royalty base. That base is usually the net receipts of the company for a product or process covered by the invention. Actually, just how the base is calculated is a very gray area and depends greatly on accounting methods. For instance, let's assume that you have invented a $10 component for a large machine which is priced at $100,000. Obviously, the royalty base is not the price of the entire machine. What is done in such a case? One approach could be to attempt to estimate the increase in the overall efficiency of the machine as a result of your invention. If the efficiency of the machine is increased, say, by 5%, the royalty base would be $100,000 x 0.05 = $5,000, which is then multiplied by the number of such machines sold. Sometimes, the royalty base may be reduced by costs of service, repair, advertising, shipment, etc. However, I do not think that is fair to the inventor, as he usually has no control over the production, service, and other such aspects of the business.

The royalty percentage may be calculated on a progressive or sliding scale to help the manufacturer recover its investment faster. For instance, the royalty may start at 3% for the first 10,000 units sold, increase to 4% for the next 50,000, and become 5% thereafter.

A very important part of the deal is the so-called *minimum royalty*, which is the minimum amount of money the inventor receives regardless of how many units are sold. This payment may force the company to be more active in reducing the invention to practice, rather than sitting on it and taking it easy, just to keep the competition at bay. Actually, such behavior is illegal, and an inventor may sue a company for obtaining the license and then not using the invention. I once knew the inventor of an electric muscle stimulator intended to reduce back pain. The Johnson & Johnson Company obtained a license from him, but did not use the invention for a long time, in order not to interfere with the promotional program for Tylenol. The inventor successfully sued the company and won millions of dollars in damages.

In concluding these observations, I will say a few words about middlemen. These people, who may act as your agents, can be helpful in raising money, finding an interested party, and even helping you to negotiate a deal. Many of them, however, are just wheeler-dealers who waste your time without delivering any results. You should deal with such people strictly on a performance basis. Any agreements with middlemen should be limited to a short time, usually no more than six months. Typical compensation is between 5 and 10% of received licensing fee and royalties received. It is important to limit such payments to several years, say two or three. This, however, can be done only if this is not an equity deal, that is, if the middleman does not receive stock for his services. If it is a stock deal, arrange for the stock to become vested only after the deal is done and the money is received in the bank. Never give middlemen shares in exchange for mere promises.

Have fun and good luck!

ORDER FORM

Additional copies of *"Adventures of an Inventor"* by J. Fraden may be purchased directly from the publisher

Price per copy: $19.95

Number of copies   ....

Shipping and handling (in U.S.), for 1st copy

    Overnight delivery: $7.50

    1st class: $5.00

    Book rate: $3.00

    each additional copy add $1.50

California residents add

sales tax of 7% $1.40

Total: $ _____

Send your check or money order (U.S. funds only) to:

Distribution Dept., *Hurricane Books Publishing Co.*

P.O. Box 927412   San Diego, CA 92192-7412